THE NEW NATURALIST

A SURVEY OF BRITISH NATURAL HISTORY

BIRDS AND MEN

The aim of this series is to interest the general reader in the wild life of Britain by recapturing the inquiring spirit of the old naturalists. The Editors believe that the natural pride of the British public in the native fauna and flora, to which must be added concern for their conservation, is best fostered by maintaining a high standard of accuracy combined with clarity of exposition in presenting the results of modern scientific research. The plants and animals are described in relation to their homes and habitats.

The text and line illustrations are here reproduced unaltered, but the process of manufacture used to achieve an economic price does not, unfortunately, do full justice to all the photographs; and those originally in colour appear in black and white.

THE NEW NATURALIST

BIRDS AND MEN

The Bird Life of British
Towns, Villages, Gardens & Farmland

by

E. M. NICHOLSON

Bloomsbury Books
London

First published in 1951
Collins, 14 St. James Place, London

This edition published 1990 by
Bloomsbury Books an imprint of
Godfrey Cave Associates Limited
42 Bloomsbury Street, London WC1B 3QJ
under license from William Collins Son's & Co. Ltd.

ISBN 1 870630 14 9

CONTENTS

* As explained on the second page, these plates are here reproduced in black and white.

* PLATES IN COLOUR

It should be noted that throughout this book Plate numbers in arabic figures refer to Colour Plates, while roman numerals are used for Black-and-White Plates

PLATES IN BLACK AND WHITE

PLATES IN BLACK AND WHITE

EDITORS' PREFACE

THE Editors of the New Naturalist would wish for no other author of the first bird book in their main series than E. M. Nicholson.

During the present century ornithology has changed from a particular study by a coterie of scientists and privileged amateurs to a much wider subject, with a considerable public following. Nicholson was one of the first to recognise this change. His own interest in birds had developed early; he grew particularly interested in bird territory and ecology through the influence of H. Eliot Howard and others, and was the first British worker seriously to embark upon a proper classification of bird habitats, and an assessment of the populations of common land-birds. His interests have included, and still include, marine ornithology, the comparative and evolutionary study of bird songs and voices, the relation of birds to their habitats—their autecology, the study of the numbers of birds and their changes: all these subjects were new when he first explored them, and particularly lend themselves to co-operative investigation by teams of collaborators.

In his earliest book, *Birds in England* (Chapman & Hall, 1926) Nicholson showed that he had come to realise that our knowledge of birds could be substantially increased by encouraging amateurs to join the ranks of the critical investigators. His second book, *How Birds Live* (Williams & Norgate, 1927) dealt purely with the habits of birds. In 1931, however, with *The Art of Bird Watching* (Witherby) he returned to the task of promoting co-operation, in order to solve purely ornithological problems. The next step was inevitable: when the British Trust for Ornithology was founded in 1933, he was its first Secretary, and a close association between organised amateur ornithology and the University of Oxford began, which soon resulted in the foundation of the Edward Grey Institute of Field Ornithology. During the rest of the 'thirties Nicholson devoted a great deal of his spare time to the Trust, though his contributions to published ornithology during this period included many scientific papers and two books on the *Songs of Wild Birds*

(Witherby, 1936 and 1937) which show that his aesthetic approach to birds is as sensitive as his powers of scientific criticism.

The second world war failed to sever Nicholson's connection with active ornithology, though he had to watch his wrynecks at Potsdam, and make notes at Cairo and Quebec, and listen to jackdaws in Downing Street. For Nicholson's *alter ego* is Secretary to the Office of the Lord President of the Council. Not long ago he became Chairman of the British Trust for Ornithology, and once more active in the organisation that he has done so much to build; now he is Chairman of its Research Committee.

Nicholson must not be thought to be in any way a victim of the machine he has helped to create. He has always seen beyond the problems of organisation to the objects for which they were devised, the extension of ornithology into fresh fields. When the editors of this book asked him for his own concept of his aim of an ornithologist his reply contained no mention of organisation. It was as follows: "To look at birds . . ., to describe faithfully and . . . to measure resourcefully and accurately, to speculate stimulatingly but tentatively, to set things in their historical and biological perspective, and to recognise that ornithology is the business and the joint contribution of all ornithologists, whoever and wherever they are".

We greet this new book; a book about the impact of man upon birds, and of birds upon man—the subject that has been close to his heart for many years, and on which he has been collecting multifarious notes. The result of his analysis and synthesis is—we know its readers will agree—a most stimulating and original contribution, and a new milestone in the progress of British ornithology. We are sure readers will be particularly delighted at the note of optimism that pervades it—the suggestion that not all the works of man are fated to destroy nature, and that some are likely to improve the quantity and variety of our birds. It is clear that we have a most interesting half-century before us, from the ornithological point of view; and at the end of it this work will, we expect, be still widely read, and quarried for facts and conclusions, by the next generation of comparative ecologists and field naturalists.

THE EDITORS

AUTHOR'S PREFACE

When the Editors invited me to write this book they offered me an opportunity to complete a picture of the impact of civilisation on our bird life for which I had drawn a first sketch in *Birds In England* over twenty years ago. Here was a chance to trace the shaping of our towns and countryside and the life-histories and ecology of their most characteristic birds. In taking this opportunity I soon found that far too little is yet known to allow anything approaching a complete picture to be drawn, although it is true that enough has been learnt recently to give us a much better idea of the subject. What I have done here, therefore, amounts to no more than a fuller and more mature sketch from a new angle, emphasising salient features and a certain amount of detail, and bringing out some of the gaps remaining to be filled in our knowledge of the facts and of underlying forces. I have not hesitated to look forward as well as back, since nothing is more misleading than to accept the assumption that the particular pattern of bird life which we see in any given place at any given time is in any way more normal or enduring than the different patterns which have just preceded and will soon follow it. For a similar reason I have tried to describe bird life in Britain against the background of the distribution and habits of the same birds in other lands, not exhaustively but as a reminder that the birds we see here also live in other countries, sometimes on different terms. To begin to understand, say, a wren, we need to take not only a wren's-eye view of the nearest heap of brushwood but a broader sweep of vision over the thin brown line of wrens which girdles the northern world from Iceland and Ireland through Europe, Siberia and North America to the Gulf of St. Lawrence.

This book is concerned with the normal habits and characteristics of a selected group of birds, and with variations and exceptions in habits which throw light on the normal. The fullness or brevity of treatment or the omission of reference to a species is determined by the degree to which it has become dependent on men in the British

Isles and has come to rely largely upon types of habitat greatly modified or actually formed by human action. In the description of species attention has similarly been concentrated on those aspects which throw most light on their capacity to adapt themselves to or to exploit civilisation. A large number of references to aberrations and isolated occurrences have been omitted to avoid cluttering up the text with detail or with relatively unimportant qualifications to statements which apply in at least ninety-nine cases out of a hundred. Where such occasional variations are unmentioned it is because they have not been considered of sufficient importance, but no doubt others would sometimes select differently, while occurrences which now seem exceptional or erratic may be shown to be more usual by fuller observation, or may represent the first appearance of new habits which will later become widespread, such as paper-tearing by tits.

It might well be argued that certain species not fully treated here, such as the chaffinch, blue tit, whitethroat and kestrel, have as good a claim as some of those described at length. This problem where to draw the line is inherent in the series of *New Naturalist* books on birds as planned at present, which aims to review British bird life in a series of broad habitat groupings; it is the price we have to pay for the greater freedom to trace the relation of birds to their environment permitted by the scheme. Birds whose main or most typical habitat does not fall within the scope of this work, such as birds most densely distributed in woodland, or on uncultivated heaths and moorland are to be fully treated in the other books.

In describing the more characteristic species within its scope this book deals more fully than usual with habitats, ecology, population, and distribution, which repay more attention than they frequently get. Other aspects are more summarily dealt with for various reasons. Identification details are amply available elsewhere, and no attempt has been made to provide here more than a general sketch of what the bird actually looks like in the field, based upon notes made in field conditions instead of descriptions taken from corpses and skins. Nests, eggs and breeding biology can also be so fully and conveniently traced elsewhere that extended treatment has not seemed necessary for them. Description of display has been kept within strict limits since it can so easily take an excessive share of space and tax the reader with visualising movements much more appropriately recorded by the lens and the film than by the eye and the written word. Bird songs and calls have been more fully described and recorded with gramophone records, by Ludwig Koch and myself in *Songs of Wild Birds* and *More Songs of Wild Birds*, and are also briefly treated here.

I have tried in each case to bring out the different character and way of life of each species, and in choosing sources I have generally preferred to use my own notes rather than to compile from the writings of others, although I am of course indebted to many other observers who have broken new ground or have dealt with aspects outside my own limited experience. It will generally be apparent from the text where I am writing from my own observation, and in most cases I have tried to indicate where the observation was made, since this is often more relevant than used to be believed when the great variety and constant evolution of bird habits were less well understood. A brief note on any important published sources used in each chapter of the text is added at the end of the book. Witherby's *Handbook of British Birds*, which has been constantly at my elbow for points of distribution, migration, food and breeding biology, is not cited elsewhere, since its essentiality as a source is self-evident.

In accordance with modern practice for field ornithology I have dealt with species, not subspecies, unless the subspecies is clearly separable in the field and its separation is significant for the bird-watcher as distinct from the taxonomist. Scientific names of the principal species treated are listed at the beginning of each of the chapters concerned, and a complete list of scientific names will be found on pages 231-35. I have used English names in accordance with the *Handbook* except in the following cases, where I consider[1] a change is called for :—

from	Hedge-Sparrow	*to*	Dunnock
	Song-Thrush		Throstle
	Great Spotted Woodpecker		Pied Woodpecker
	Lesser Spotted Woodpecker		Barred Woodpecker
	Great Black-backed Gull		Great Blackback
	Lesser Black-backed Gull		Lesser Blackback
	Common Gull		Mew Gull

[1] See *Ibis, 91:* 692 (1949).

In all these cases the alternative name is already widely used, and in some it is the older as well as the briefer and more convenient English name.

Finally let me apologise to the readers of this book for its many shortcomings and hope that some of them may feel moved to help in dispelling some of the cloud of ignorance which still covers so much of its subject. Those who do so will not be disappointed. No part of ornithology is more interesting, no part is seeing more rapid change, and none is more important.

E.M.N.

LONDON
25 June 1950

CHAPTER I

HOW MEN HAVE SHAPED NATURE

THIS book will aim to tell the story of how men have shaped nature in Britain and have thus cast the bird population into its present mould. It will outline the ways in which the affairs of birds and men have become mixed up together, and will describe the birds of farmlands, of garden, orchard, and hedgerow, and of towns and buildings, leaving to other writers in this series the description of sea and shore birds, of marsh and inland water birds, of mountain and moorland birds and of the birds of woodlands and forests. Finally it will discuss some possible ways of enabling birds and men to live more harmoniously together in Britain. The ground to be covered is vast, and could not be covered thoroughly in one book, even if the subject had been sufficiently explored by research, as it has not. If inevitably this book must fail to provide a comprehensive and final statement over its entire field I hope at least that it will not fail to show how important and how interesting that field is, and to stimulate further and more pointed studies.

To shape a landscape is to shape its bird life. The occurrence and scale of mountains, hills and moorlands, of forests and trees, of lakes, rivers and marshes, of cliffs, shingles and mudflats determine so definitely the general character of bird populations that an expert ornithologist with a good map could make a useful guess at the birds likely to be found even in an entirely unfamiliar area. This book however is not concerned with the geological, climatic and other natural forces which have fundamentally moulded the British landscape; it takes up the story only at the point where man comes on the scene.

The landscape of Britain now is a work of nature more or less modified and in extreme cases practically obliterated and superseded by man's efforts. These efforts, although so complex in their effects, have primarily been achieved through one or other form of five elementary instruments—the axe, the spade, the plough, the hod and the wheel.

Man's first foothold in Britain, in contrast to what happened in North America, was won before these instruments had been perfected and become widely available. For a long time, therefore, most of the areas in which human population is now concentrated remained practically inaccessible, covered by vast masses of trees or inland water, or some impenetrable blend of both. The birds which lived in these forests and swamps gained protection from human pursuit owing to the dangerous wild beasts, such as wolves, which still found shelter in them, while men were powerless either to clear the forests and drain the swamps or even to exploit them at all extensively. Probably only the upland areas, some of the more barren soils and parts of the coasts and outlying islands were freely accessible, and these were too exposed or too inhospitable to support more than scattered inhabitants.

In the first long phase of human colonisation, therefore, it seems likely that forest and marsh birds continued in occupation of an immensely larger part of Britain than anything we have known within historic times. There are now less than 3 million acres out of Great Britain's 56 millions under any sort of woodland; before the axe got to work there must have been at least 20 millions, and perhaps substantially more. The acreage of more or less unwooded marsh and fen, particularly in eastern England, must have been relatively much greater still in comparison with the present miserable remnants. But the scale and character of the hill grazings and moorlands must have changed very much less, although burning and grazing have destroyed much of the primitive texture. They can never have been densely occupied or much cut about and altered by man, and in so far as they were occupied the human population has been retreating from them for many generations. We know that as an exception the chalk downs of southern England were quite considerably populated in neolithic times, and these downs with their camps, lynchets and barrows clearly bear the aspect of places long inhabited by people but now again largely given over to wild life. Many of these places, such as that where Stane Street crosses the Sussex Downs, or the Great Ridge Wood on Salisbury Plain, are now among the loneliest and most primeval-looking in southern Britain, while others which until far later times were virgin forest and untrodden marsh are now overrun by men and their buildings.

Field birds such as skylarks and lapwings, birds attached to livestock such as swallows, and sharp-eyed opportunists such as jackdaws and other crows must have been among the first species to adapt themselves to man as a herdsman and later a cultivator of the more open and less unfriendly tracts of the uplands and elsewhere. Unquestionably the relations which go furthest back and have least changed must be those between birds and man as a farmer. Man as a hunter has no doubt been active for even longer. Avebury Museum contains the femur of a "blackbird" (the species cannot actually be identified) which was found with, and is capable of reproducing ornamentation on, clay pottery dated about 2500-1900 B.C. Here man the hunter is already supplying man the craftsman with tools, and man the farmer has just begun to remould Britain. But many of the primeval quarry have probably become extinct and it is unlikely that many of the present sporting birds were effectively attacked until comparatively modern times, especially since a surprising number of them are now represented by either wholly or largely introduced stocks or by quite recent colonists. Except in giving birds grounds to fear him man the hunter has left little impression on bird life compared with man the farmer.

As Sir John Clapham put it in *A Concise Economic History of Britain from the Earliest Times to* 1750 :—

> " . . . . . when the first farmers crossed a narrower Channel . . . . . Lowland Britain was . . . covered with 'damp oak' forest—oak and hazel, bramble, and thorn, thorn, thorn. Valleys were waterlogged and marshy. Great parts of the drier ground were also stocked by nature with well-grown oak and ash, beech, birch and yew—hard to clear with stone axes. Forest or scrub probably spread over the dry chalk and limestone ridges—the Downs, the Cotswolds and such. But the vegetation was more open and had some natural gaps. These you could extend and your goats would nibble down the seedlings; there is nothing like a goat for stopping the spread or revival of forest. In any case it is on such ridges, from the South Downs westward into Devon, that ten or a dozen settlements of these first farmers have been examined. They were mainly cattle keepers; but they also kept goats, sheep and pigs, and they cultivated little plots of wheat, probably using some sort of digging stick to turn the ground, or a pick made of a deer's antler to furrow it."

These first farmers only got to work about 2400 B.C., and not much more than a hundred and forty generations of countrymen, or about 4300 generations of skylarks, have intervened between the first faint contact between birds and farmers in Britain and the tractor-driven agriculture of the present time. During the greater part of this period British agriculture was an almost negligible influence, the increase of population and of land under the plough in lowland Britain having only seriously begun about the 8th century B.C., while only in the 1st century B.C. did the total amount of land worked with oxen reach an appreciable scale, and it was only in that century also that agriculture began to produce a sufficient surplus to enable the beginnings of the first towns to emerge.

It was probably not until the Roman invasion that man even began to emerge as a really important influence on bird life in Britain. The axe then got to work making the first sizeable clearings in the forests, and paved roads constructed by engineers enabled the wheel to begin its task of spreading civilisation. While it is impossible to trace the effects on bird-life of this process of clearing the forests, which continued off and on for at least fifteen centuries, there is no reason to suppose that it was on balance adverse either to the variety or even to the total numbers of woodland birds. A high proportion of the bird population of woodlands in any case frequents the fringes, and most of the characteristic woodland species seem to have adapted themselves without difficulty to parkland, orchards, hedgerows and gardens, where their densities are probably sufficiently increased to compensate for a large part of the vast reduction in forest area. Some species which have not found it easy to adapt themselves must have suffered considerably—willow-tits and marsh-tits, siskins and redpolls, which all delighted in the alder and willow swamps, and probably the woodpeckers, which appear only now to be finding fresh niches. The only known woodland breeding species which may have inhabited the old forests and become extinct on account of clearings is the goshawk, and the evidence that it ever was a regular British breeding bird is flimsy. The honey-buzzard must have been handicapped too, although its actual extinction in recent times must be attributed to selfish egg-collectors and ignorant game-preservers rather than to the woodman's axe.

Comparison of British woodland and garden birds with those of the Continent confirms that, serious as the vast destruction of trees has been in other ways, it has not proved anything like such an impoverishing influence in the variety of our bird life as the outlying geographical

PLATE I

Eric Hosking

Coal tit leaving nest in petrol tin

T. M. Fowler

Stock-dove

Eric Hosking

Curlew and newly-hatched chick; Yorkshire Pennines

PLATE 4

S. E. Palmar

Hooded crow at nest with young; Oban

position of Britain and its short and unreliable summers. But if comparison is limited to density of total woodland bird population in relation to area, Britain probably compares favourably with most parts of Europe, and also of North America.

While the wielding of the axe and the invasion of the wheel and the plough have probably not injured our arboreal birds to anything like the extent that might have been expected, they have carved out great new habitats in which large numbers of field and town birds have been able to establish themselves within historical times. The restless teeth and feet of sheep, goats, cattle, pigs, horses and other animals kept by men have greatly assisted the replacement of woodland and heath by fields. There can be no doubt that the net effect has been a big increase in total bird population. Any feelings of comfort which may be drawn from a review of the effects of woodland clearance are however soon shattered when we turn to consider the disastrous results to bird life of the drainage of the marshes. Historically, this process followed the felling of the forests without much overlap, and while much damage to bird life was no doubt inflicted during the medieval period of gradual human infiltration into the marshlands the worst has been concentrated probably during the past five hundred years, when records enable some judgment of it to be formed. During that period we have seen the extinction as British breeding birds of the crane and spoonbill, followed in the nineteenth century by the avocet, black-tailed godwit, Savi's warbler, black tern and ruff and reeve. These are merely the authentically recorded losses, and it is quite possible that other marsh-birds were wiped out unrecorded in the same process.[1] Others such as the marsh-harrier and the bearded tit

[1] Some of these marsh birds have lately been encouraged by both accidental and deliberate human assistance to start breeding again, especially where there has been a return to something approaching natural marsh conditions. Black-tailed godwits have bred sporadically, black terns formed a small and ephemeral breeding colony in Sussex in 1941 and 1942, and avocets have successfully bred in some numbers in Suffolk since 1947. There have also been many extended visits by spoonbills and ruffs, and in 1945 two pairs of black-winged stilts bred at Nottingham Sewage Farm, this being the first known record. Another small marsh bird, the moustached warbler, bred at Cambridge Sewage Farm in 1946, also for the first time on record. The bittern, which resumed breeding in Norfolk in 1911 after being missing for nearly half-a-century is now spreading far beyond East Anglia. Such windfalls of good fortune have as yet no firm prospect of continuance, and it remains to be seen whether they represent anything more than a passing interlude in the centuries-old process of losing our marsh-birds. They do, however, prove that the trend can be reversed, if we are determined enough to reverse it.

have been reduced to such cramped remnants that their survival is in doubt almost from year to year; it looks as if the recovery of the breeding stock of bearded tits since the blizzards of early 1947 has been largely due to a fortunate influx from the Continent. The rich group of habitats formed by British marshes and fenlands has been ruined by wholesale and ruthless drainage. The loss is the more regrettable and unaccountable because the Dutch with their passion for and knowledge of marsh drainage have yet managed to preserve intact considerably more adequate fragments of original fen.

Up to the present time the birds of Britain have seen men first prowl as scattered feeble hunters, along the coasts and estuaries, next colonise and farm the more hospitable and less exposed uplands, and then in increasing numbers invade the great lowlands, clearing away the forests, draining the fens, creating much rich farmland and in recent times burrowing into or stripping off the surface locally for minerals and covering more and more of it with artificial and often sterile new habitats of brick, stone, concrete, tarmac and steel. Until recently man's most drastic interventions could not go beyond limited modifications and extensions of existing habitats. The cleared woodlands and drained fens gave place to fields which merely spread over the lowlands something not entirely unlike the open grasslands naturally characteristic of other parts of Britain, or to gardens and orchards which were simply man-made arrangements of trees and shrubs. The fields were divided by hedges and windbreaks and banks using the stuff of nature, and the fens by straight dykes of water instead of naturally chaotic watercourses. Even the buildings were usually of local stone or timber or clay and the highways of the local dirt. It meant no serious strain for the bird population to adapt itself to these gradual and limited changes.

The recent spread of a dead skin of brick and asphalt and concrete and slag roadways and runways and works of engineering and construction have set the birds a new and much tougher problem. This problem might well prove insoluble but for the hopeful element that what makes these things intolerable for birds ultimately makes them also intolerable for men. Town and country planning, landscaping, afforestation, provision of amenities, and carelessly dropped high explosives are modern civilisation's ways of arranging to mitigate its own sterility and ugliness by enabling nature to come back to the very places where it has most determinedly been destroyed.

Apart from the trends which are already plainly visible further threats are impending. Higher standards of living and greater mobility are permitting the human population to flood and overrun areas previously little disturbed—there is very little peace in summer now for shorebirds breeding round most of the British coasts, for instance. The growing pressure of competing claims to the use of land, and the development of such inventions as the bulldozer, the tractor, tracked vehicles, and excavators are making it economically and technically feasible to invade and destroy or modify habitats so far immune owing to their remoteness, steepness or roughness, or to the fact that they were not worth while to work with more primitive tools. Agriculture, forestry, hydro-electric and water supply undertakings and the Defence Services are all reaching out into areas previously immune from more than the most superficial human attention and interference, and in another generation the fringe of natural unmodified habitats which still surrounds the humanly exploited core of Britain may well have been whittled down to a fraction of what now remains, unless much more effective conservation policies are rapidly applied.

So far the broad effect of civilisation has been considered in relation to some of the main habitats separately, but that is not the whole story. Bird populations are not static patterns arranged strictly in accordance with the various habitats. They are dynamic and respond to the principle that nothing succeeds like success. When therefore largish areas are modified and broken up into a patchwork of different and discordant habitats the result is not necessarily to create corresponding islands and strips of the appropriate specialist bird communities for each. It may instead be the utter rout of the more specialised and interesting birds and their wholesale replacement by a few aggressive, dominant, adaptable all-rounders. Even a big increase of bird population may be a misfortune if it consists of say starlings, sparrows, rooks and gulls accounting for the whole of the expansion, and perhaps on the strength of it crowding out some of the woodpeckers, house-martins, lapwings and curlews from places which they might otherwise have held on to.

The problem of maintaining a rich and diversified bird life is not just one of enabling each species to continue to enjoy conditions in which it can hold its own, but of seeing that each species continues to have a broad and strong enough base on which it can successfully compete. It is the birds whose competitive bases have become too weak or narrow—the choughs, the bearded tits, the kites, the Dartford warblers and even the wrynecks, the hobbies and the barn-owls—which form the greatest problems for bird protection. On the other hand species with perhaps equally small absolute numbers but an inherently stronger competitive position, such as the black redstart, the little ringed plover, the kingfisher and the great crested grebe, do not give rise to the same anxiety, because they alone appear to be able to exploit certain habitats which are widespread or increasing, without much risk of being repressed by more dominant forms.

With this general introduction it is now time to consider some of the main modifications which man has imposed on the natural environment in Britain, and the response which various species of birds have made to these challenges and opportunities.

CHAPTER 2

BIRDS OF TOWN AND FARM

In the course of generations, but at an ever more rapid pace, birds have become dependent directly or indirectly on people in Britain in many and varied ways. In this chapter an attempt will be made to unravel some of the main threads and to show how human arrangements for shelter, for transport and services, for food-getting, defence, sport and other activities have become the basis on which much of our bird population now exists.

In order to provide shelter for living, people in Britain have covered a sizeable fraction of the more fertile and less exposed parts of the country with dwellings of various kinds. Wattle-and-daub, stone, brick, timber and now even concrete, steel and aluminium dwellings, ranging from small bungalows and cottages to houses of three or more storeys, terraces and rows, mansions, blocks of flats and hotels have blotted out many different habitats and created an ecological vacuum into which the most adaptable birds have poured, making use of the new opportunities.

The most obvious use for birds to make of these humanly provided substitutes for rock and crags is simply to sit on them, and all over the country we see birds sitting on houses, either to take a rest, or to exploit them as convenient singing stands. There are two birds—the starling and the black redstart—which seem to prefer a chimney-pot or a commanding roof-ridge to any other song-post. Even such an arboreal species as the blackbird switches readily from treetop to housetop, and in Chelsea where I live blackbirds seem to be more and more preferring the artificial to the natural perch.

Other birds are fond of sitting on roofs when they have nothing else to do. In some seaside towns gulls are conspicuous roof-sitters, particularly herring-gulls, which like to occupy a chimney to themselves if possible. At Padstow I have seen even an adult great blackback perched on a roof facing the harbour. This habit seems most frequent on houses immediately facing the sea, although at Trebetherick, also in Cornwall, I have seen herring-gulls perched on roofs over a quarter of a mile inland. By January 1950 great blackbacks had taken to roof-sitting, among herring-gulls, by the Thames in Inner London. It is surprising that such adaptable birds as rooks, which live so close to man, should so infrequently perch on buildings, although the shyer carrion-crow quite often does, at any rate in London, and of course the crag-loving jackdaw takes to them as a matter of course.

After the young are fledged martins have a special liking for sunning themselves on suitable warm roofs. At Newmarket on the hot afternoon of 9 August 1947, I saw parties of young house-martins clustered on several adjoining roofs, picking in each case the exposure which gave them the full benefit of the sun, in spite of the fact that the shade temperature at the time must have been around 80 degrees Fahrenheit. In August also I have seen somewhere about a hundred house-martins —a notable assembly for this species—sunning themselves every morning on a roof in Hertfordshire.

Beneath the roof, and before coming down to the rooms occupied by people, there are chimneys and attics and rafters and eaves which are used for roosting and breeding, particularly by the swift and the starling, for both of whom they form one of the favourite nesting sites, and by the house-sparrow which is more ready to choose a position giving directly on to the open air. Paradoxically there are in practice few nesting sites so immune from human disturbance as those under the roofs of dwellings, and the nest of the swift, which is so often in this position, must be less often seen than that of any equally common bird.

Although the swallow nests in chimneys in some districts probably the only wild bird which does so in Britain in large numbers is the jackdaw. In some places the jackdaw's fondness for nesting in chimneys, which it needs to block with sticks for the purpose, leads to annual warfare, and in a Wiltshire village I was woken up one Sunday morning early by what sounded like an outbreak of hostilities, but proved to be merely the inhabitants armed with shotguns trying to protect their homes from invasion.

The outside of dwellings, immediately under the eaves, is the

favourite nesting site of the house-martin, which in this respect is parasitic upon man and is itself parasitised in turn by house-sparrows. These find in the martin's nest a more suitable site for their own than almost any which man provides in modern buildings.

Many dwellings are partly covered with Virginia creeper, ivy or other climbing plants or trees, and these offer some attractive sites both for roosting and nesting. House-sparrows are very fond of them for both purposes, partly because they remain arboreal birds at heart and are never wholly satisfied with brick and stone, and partly no doubt because they give the most nearly cat-proof comfortable cover to be found in many towns. During the war when the number of cats in London was much reduced I trained some of the creeper *Polygonum baldschuanicum* up my house in Chelsea and it promptly became a fair-sized sparrow-roost for the neighbourhood. Blackbirds then moved in and reared a brood in a nest about 20 ft. up in the creeper; but this did not save at any rate one of them from being caught by a cat immediately after fledging. On a house on the edge of Birmingham, also in a cat-infested district, I was shown in June 1947 a very tame throstle sitting on 4 eggs in a nest (from which a previous brood had lately flown) about 12 ft. up in a fruit-tree trained up the wall, and just beneath a bedroom window. In some cases at least drainpipes serve as substitutes for creepers, and in April 1946 a pair of blackbirds built a completely cat-proof nest about 20 ft. up on a house opposite mine in Chelsea between the wall and a crook in a drainpipe, protected from above by broad overhanging eaves, without any plant support or shelter at all.

Probably the most characteristic and persistent of all creeper-nesters, in suburban and country districts, is the spotted flycatcher. Although a nervous bird, with little trust in man and a well-justified dislike of cats, the spotted flycatcher seems unable to resist building on a well-trained espalier or creeper, not too densely grown, preferably just out of arm's reach, and on the sunny side of the house itself more often than on a garden wall. Blue and great tits, pied wagtails, wrens, redstarts, pied flycatchers and other birds sometimes nest in holes in walls of dwellings and outbuildings, or in the thatch where there is one.

Thatch is the only part of human dwellings which is habitually attacked by birds, pulling out the straws or reeds in order to get at insects hidden inside. Great tits are particularly complained of on this account.

Swallows often nest in porches and outbuildings where the doors or windows are usually open, and I know a house where swallows habitually flew through the hall and the sitting-room into the herbaceous garden, but birds do not normally make use of the inhabited rooms of houses. A pair of robins at Birmingham, however, started to build their nest in an unmade bed during the breakfast hour. When the owner realized what was happening the bed was left unmade, and the birds eventually reared a family in it. This fortunately is not something that people often have to put up with, although there have been many reports recently of blue tits and other species getting a habit of entering dwellings and even roosting inside them. The apparent efforts sometimes made by robins, pied wagtails and other birds to batter their way through closed windows are not really efforts to get into the house but attacks on the reflection of the bird, which is mistaken for a rival male.

Except while food-rationing intervened, more and more people have put out food for birds, especially in winter when they need it, and this has led to such a strange sight as black-headed gulls and wood-pigeons alighting on window-sills of blocks of flats or resting on nearby parapets, as well as great, blue, coal and sometimes other tits and nuthatches hanging on coconuts or suet, and robins, blackbirds, starlings, sparrows and other garden species jostling on birdtables or coming to drink or bathe at birdbaths.

Birds, unlike certain mammals and insects, rarely help themselves to any processed human food which is not meant for them. Great and blue tits have however developed in a number of areas a habit of pecking through the lids of milkbottles in order to drink the cream. These birds seem to have an insatiable craving for fats, possibly connected with their small size and incessant activity, and during the wartime hard winters they must have particularly felt the loss of the suet and bacon-rinds and similar food which was so widely provided for them up to 1940. Loaves of bread left out on doorsteps by the baker are also liable to be regarded as fair game by house-sparrows.

Large blocks of flats and hotels form a somewhat different habitat from ordinary dwellings and in America particular bird habits have already been adapted to them. For instance in Washington D.C. the "turkey-buzzards" or turkey vultures resting lethargically on the roof of the Shoreham Hotel and of apartment buildings on Connecticut Avenue are a familiar sight, and the American nightjar or eastern nighthawk has taken to nesting freely on the pebbly flat roofs of some buildings. In Britain adaptations of this sort are not plainly evident yet. An interesting partial adaptation is the use by blackbirds and other garden species of high roof-gardens such as that on Derry and Toms' shop in Kensington. A St. James's Square house-sparrow not long ago formed the habit of flying into the main conference hall at the Ministry of Labour and going to roost on a ledge above the Minister's head during important meetings.

Churches and large public buildings with spires, towers and colonnades also provide opportunities which have not yet been fully seized by birds in this country, although in London there are some spectacular starling roosts on St. Paul's, the National Gallery and the Nelson Column, for example. The peregrine has probably occasionally and the kestrel more frequently bred in spires or towers—a pair of kestrels once nested on the Nelson Column. Both species however seem to haunt large buildings in towns less in Britain than on the Continent, where, before the war I have seen as many as five kestrels playing about Cologne Cathedral spires (in April), and young learning to fly around the Frauenkirche at Munich, while others were using such great buildings as Ulm Minster and the Duomo at Milan besides many in other towns. The peregrine I have seen in winter round buildings in the middle of Cologne and of Montreal (where it has been reported as breeding on the Sun Life Buildings), and one has recently been seen on the top of a large building in central London. I have seen, in October, a tiercel circling over St. James's Park lake. Another bird which, although it has colonised some of our towns, appears less attached to large public buildings here than abroad is the black redstart. The recent colonisation wave in Greater London however began from a sort of public building—the Palace of Engineering at Wembley, and there have been nests within the precincts of Westminster Abbey.

Rooks, which so rarely perch on buildings in England, do so freely in parts of Ireland. At Dunlaoghaire, which we used to call Kingstown, I have seen a church roof-ridge carrying an iron grille decorated with rooks and jackdaws on the pinnacles and starlings all along between. At Bray, Co. Wicklow, in September 1927 I watched a very obvious and interesting case of birds using the roof of a large hotel for play. A rook kept settling on top of the flagstaff, with some trouble owing to the strong wind, and the jackdaw then swooped above, doing everything possible to dislodge him. If he failed after half-a-dozen assaults the jackdaw gave it up and dropped back onto one of the chimney-stacks, whereupon the rook, unwilling to give up the game, took wing and evicted him from his chimney. A rush to get first to the flagstaff followed. If the jackdaw got there first he was never able to hold the position; if the rook reached it first he was often prevented from alighting by being so closely pursued in the stiff wind, but once down he could not be evicted. A cruising immature gull—a common or mew gull, I believe—on seeing the game twice intervened to prevent the rook from getting his perch, but took no further part.

Factories and industrial buildings, in spite of the area they occupy and the recent movement towards making them more attractive, appear to be remarkably barren of bird life in Britain. The black redstart at Wembley and Woolwich and elsewhere has shown some signs of becoming a factory bird, but for the most part industrial buildings offer neither suitable cover nor food, and unless they form suitable resting places, say for gulls from an adjoining piece of water, they are liable to be a total loss for the birdwatcher. Perhaps, now that we have lily ponds at certain pitheads we may some day see successful experiments in placing nesting-boxes round those factories whose surroundings offer a living to small birds. Certain types of industrial building have special attractions or handicaps. The warm cooling towers of electricity generating stations are in some cases favoured by starlings for winter roosting. Cement works with their pervasive whitish dust do not seem to be as repellent to birds as might be expected; there is one near the Western Region main line west of Newton Abbot which has for years smothered the birds, nests and trees of a neighbouring rookery with dust without causing the rooks to desert. Abandoned mines of various sorts with their open shafts and ruined buildings are attractive to various species. Opencast mining, like quarrying and the excavation of clay, sand, chalk and gravel may give rise to suitable conditions for such infrequent or rare species as the sand-martin, wheatear, stockdove and even little ringed plover. The little ringed plover has apparently begun to colonise south-east England in places where the topsoil has been stripped and standing water created by excavations for building purposes or for reservoirs. Such habitats are of course created at the expense of the pre-existing bird population, which rarely receives much thought, although the late H. Eliot Howard, who was a director of Stewarts and Lloyds, once told me how anxiously he examined the site of the great new Corby steelworks in Northamptonshire to determine how many bunting and other territories would be blotted out, at the time when the decision to go ahead with the plant was being taken.

Railway stations, covered markets, warehouses and other large buildings where food is handled or livestock are brought are considerably more attractive than ordinary industrial plants. I have seen rooks haunting cattle-pens in Newmarket station yard, perhaps partly for water and partly for food-leavings, and others foraging on the Great Western main line at Slough. At Truro I have seen jackdaws in search of food alighting on the station platform in the early morning, while at Portleven near the Lizard they were tamely feeding in the streets. At Castle Douglas station in Kirkcudbright in July 1947 I noticed black-headed gulls still in their breeding hoods sitting on the slate roof of a warehouse by the line. House-sparrows commonly make themselves at home in covered markets where there is an easy exit.

Transport and communications offer plenty of other opportunities for birds. Telegraph and telephone wires have long been favoured for congregating flocks and parties, or as resting-places, singing stands and vantage points for watching for prey. More lately the high-tension lines of the electricity grid and local low-tension cables have served a similar purpose. Linnets, goldfinches, yellowhammers, swallows, martins, starlings, jackdaws, rooks and turtledoves are among the birds which have taken most generally to congregating on the overhead wires; corn-buntings and yellowhammers use them freely as singing stands, and I have even seen a whitethroat at Staines singing away at the top of a naked grid pylon at least 40 ft. up. In Britain kestrels and little owls seem to be the birds most given to using the wires as an aid to locating their prey.

Accidents to birds through causing electric short-circuits have not been widely reported in England. In Yorkshire, however, electricity breakdowns have been attributed to kestrels in the early morning closely approaching pylons and causing electric short-circuits. In some cases the kestrels have been found at the foot of the pylon, badly burned across the wings. The courtship of the Indian roller has been stated to cause havoc to itself and to the distribution of power in parts of India owing to the unfortunate habit of a pair of birds sitting on adjoining high-tension lines seeking to touch bills at the climax of their performance—a climax which in this situation can never be repeated. Accidents through flying blindly into the wires appear to have decreased as the obstructions have become more familiar, but certain birds which fly rapidly and low and are not easily able to change course are especially vulnerable, most of all when they travel at night.

The Post Office have reported trouble with both green and pied woodpeckers boring holes of varying depths in telephone poles of apparently sound timber. Damping of the vibration of the wires to silence humming is stated to have put an end to the attacks, which Post Office engineers have suggested may be caused by the woodpeckers being misled into supposing that there is a bee's nest in the wood. The electrocution of the kestrels in Yorkshire has similarly been suggested by the engineers to have been induced by the birds trying to get at the source of a loud humming noise caused at times by the pylon. Full ornithological investigation of these incidents would be interesting.

Canals, reservoirs and sewage-farms, which are becoming of outstanding importance for our bird life, though man-made works are best considered as artificial rivers, lakes and marshes, and are accordingly outside the scope of this volume. Reference must however be made to the striking agility shown by parties of starlings in getting food on the filter-beds of sewage-farms while flying up every few seconds to avoid a wetting from the slowly revolving arms of the sprinklers.

Quays, piers, and harbours play an important part in providing resting places and shelter for birds, and are often also excellent sources of food owing to the waste from fish and other supplies landed, from ships' galleys, from sewage outfalls and from mud and other conditions created artificially. The gull population has largely increased in modern times and much of the increase must be attributed to the ease with which gulls can pick up a living at and round ports and seaside towns whenever they find the slightest difficulty in getting food elsewhere. Even at a little fishing village in South Devon, Beesands, I have found as many as five or six hundred gulls, nearly all herring-gulls, concentrated in October at the height of the fishing season. While the adults, which predominate, move over to feed on the farms inland wherever conditions are favourable, and compete successfully with rooks, jackdaws and starlings, the less experienced juveniles and immatures remain more brazenly parasitic on people. They hang around the fishermen's boats, eating or trying to eat almost anything

PLATE I.

Eric Hosking

Herring-gull calling

Dr. M. S. Wood

(*a*) House-martin bringing food

H. M. Stone

(*b*) Pair of jackdaws at nest in a church tower

within reach—I saw one mount guard over some vegetable refuse and drive off other immatures while he tried to devour grease-proof paper, while another attempted to eat a matchbox and some rope. It was very noticeable that the adults, being stronger and craftier, were the first to satisfy themselves and to join the gorged flocks resting on the shore. By about 11 a.m. on one morning there were 336 adults and 82 immatures in the largest of three flocks on shore, the adults thus forming over 75 per cent.; while a party still waiting for food contained 20 immatures and only 5 adults, or 20 per cent., and a party actually getting food soon afterwards was composed of 24 immatures and 15 adults. This is a subject which would repay further study. If as seems likely direct parasitism on man enables a larger proportion of such a species as the herring-gull to reach maturity—a process which takes four years—the greatest biological handicap of the species is to that extent overcome and it gains a most valuable competitive advantage as against other seabirds and landbirds also.

In parasitism on the seafarer the prizes go to the birds which are bold and enterprising enough to snatch them as they fall from the galley porthole or the fishermen's nets, and natural wariness therefore becomes a serious handicap. The bird has to learn to know when and where to keep well beyond gunshot, and when and where to eat practically out of man's hand. The more confident forms, especially the black-headed and herring-gulls, make the most successful direct parasites. But the less confident—particularly the mew gull and the great blackback—make good some of the disadvantage which their greater shyness imposes, by hanging about at the rear of the scrum and behaving like skuas in parasitising the more enterprising birds which have dashed away with the most attractive prizes. The confident species also act as decoys, their immunity lulling the fears of the more wary, so that even the great blackback is rapidly becoming in some areas a relatively tame and approachable bird. During the last few years there has been a conspicuous increase in the visits to central London of the great blackback, which was recorded there only four times in the whole of the nineteenth century, became a rare but annual visitant just over twenty years ago, and is now to be seen daily. On 29 January 1950, E. R. Parrinder, C. B. Ashby and I counted 235 on the Thames between Barnes and Woolwich. In some harbours, such as St. Peter Port, yachtsmen are seriously plagued by herring-gulls and other species availing themselves of the decks and masts of moored small craft, and leaving them in a foul state. Skuas more rarely enter harbours, although during a gale in September 1946 I saw one momentarily fly over the harbour wall at Brixham in Devon. Gannets seem to come inshore and get mixed up with fishermen chiefly when after the same fish, although a very interesting film made by G. T. Kay showed large numbers of gannets diving in Lerwick harbour in Shetland for food thrown overboard from ships and quays. Terns, again, usually remain independent, although at Godthaab in Greenland I have seen Arctic terns pelted with stones by the inhabitants as they dived into the harbour for scraps of fish which had been thrown in for them.

Waders quite often come into harbours, but apparently in pursuit of their usual food. At Portleven in Cornwall I have seen a turnstone busy under the propellers of a fishing boat lying on the bottom at low tide, and at Hugh Town harbour in the Scillies I have seen a mixed flock of turnstones and purple sandpipers running about literally underneath the public bar of the Atlantic Hotel, which is built out on stilts above the tides. At the Bishop Rock in March 1946 I saw a purple sandpiper taking refuge at the base of the lighthouse. It is, however, beyond the scope of this book to discuss the part played by lighthouses and by ships at sea as refuges and as a factor in bird migration and distribution.

Returning to the land, highways have for some years been becoming less of a helpful and more of an adverse influence on birds owing to the replacement of horse traffic by motor vehicles. Unlike the horse the motor-car provides no spillings from the nosebag or edible droppings, and with its great speed and acceleration it is a menace to birds feeding on the highway, particularly where traffic flows at above 35 miles an hour—up to which rate most birds can take evasive action even at the last moment. Worse still, the motorist demands a surface wholly barren of life and shelter, and one whose effluents are liable after rain to kill off insect and other food in the streams which it crosses. Traffic needs also lead to drastic cutting back of roadside trees and hedges, thus depriving birds of shelter. As landscaping and the planting of trees and shrubs proceeds and parkways are developed new habitats favourable to bird life may be created, but recently the birds most affected by highway conditions have had a difficult time. Even ribbon development of wayside cottages and bungalows, while it has increased the acreage of gardens and the numbers of garden birds, has also created chains of bases from which guerilla armies of destructive cats deploy over wide areas of countryside formerly beyond the reach of their claws.

Modern functional design in civil engineering offers little shelter to birds. The old bridge with its ornamental additions and its clothing of moss, ferns and so forth, was a favourite nesting-place for dippers and other hole-nesting species, and for house-martins, which used for example to have a considerable colony at Swinford Bridge on the main road out of Oxford to South Wales, and another smaller one under the arches at Newbridge, the next one upstream. Modern structures do not usually lend themselves to these amenities. Kestrels have, however, begun breeding in a severely functional concrete outlet tower built out from the embankment of a new reservoir near London.

Much the same is true of such street furniture as lamp-posts. The old gaslamps in the Green Park, for instance, offer safe, warm and attractive accommodation for roosting and breeding sparrows and I have seen a great tit prospecting one of them in the nesting season. Blue tits commonly nest in lamp-posts.

The newest form of transport, air travel, seems very unlikely to give birds any appreciable provision of either food or shelter, while the growing tendency to lay down more and longer and wider concrete runways in place of grass is as unsatisfactory for bird life as for agriculture. The need for keeping fairly large tracts of land under very short herbage is favourable to certain species, especially lapwings and starlings, but the presence of flocks of these birds liable to take wing and come into collision with aircraft is a worry to the airfield control, and flocks of largish birds round airfields are unlikely to be encouraged by the authorities. The effects of air lighthouses and beacons, especially on migration, remain to be studied; Sir Cyril Hurcomb drew my attention in Washington D.C. to the concentration of passage migrants in early mornings in autumn near the foot of the tall Washington Monument, which carries an aerial light beacon, and it is possible that development of extra powerful fog-piercing lights and flarepaths may attract migrants to halt in thick weather.

Defence works, which since 1940 have come to occupy an appreciable amount of land, cause complicated and by no means wholly adverse effects on bird life. The actual buildings, magazines, gun emplacements and so forth are probably not very important either way, although they no doubt encroach seriously in certain places on the previous territories of rare birds which are in no position to cope with further handicaps. Round the installation themselves there is usually a neglected area given up to rough herbage, thistles and weeds. This is often attractive to goldfinches, linnets, stonechats and other birds penalised by good cultivation; during the war goldfinches were even brought by this means to begin to colonise inner London, and in 1945 I saw as many as twelve together on the anti-aircraft rocket site just north of the Hyde Park bandstand, as well as two almost in Piccadilly at the Ritz Hotel corner of the Green Park.

The barbed wire coils and entanglements put up for defences, although barren in themselves, also encourage rank growth, and in Suffolk I saw at one point a red-backed shrike sitting on the barbed wire and a red-legged partridge running out from under another coil nearby, while further along the coast at intervals about half-a-dozen migrating young cuckoos were using the tubular steel beach-landing obstructions as a convenient perch while pausing on their journey.

Certain species, notably the Dartford warbler in Surrey and Hampshire, have the misfortune to find their last strongholds on heaths and commons whose vegetation the military are making a determined bid to wipe out by the mass deployment of tracked and other heavy vehicles, and certain other species are exposed to alarming risks by the impact or blast of high explosives. In some cases there is an offset in the form of greater immunity from other human disturbance and interference, but the repercussions are so complex that it is impossible to generalise. For example where grazing by sheep and other livestock is made impossible, as it often is, some of the coarser vegetation is liable to run riot and to cause unexpected and often adverse changes in the habitat. I have walked more than two miles over an area in south-west Scotland, well-known for its wild life, after some years of use as a target practice site and found only a couple of carrion-crows, a few skylarks, pipits, and mallards, one pair of wheatears, a pheasant, a blackbird and a number of black-headed gulls on ground which seemed made to carry an immensely greater and more interesting bird population; botanists had the same experience with the plants, although in both cases there was a certain amount of interest around the fringe. There are of course certain special defence works which are highly serviceable for birds after their military use has ceased; on one such erection in the mouth of the Thames I have counted as many as 58 great blackbacks, almost all adults, and 47 cormorants in

PLATE 5

E. H. Ware

Pair of chaffinches at nest with young in apple tree

PLATE 6

Eric Hosking

Turtle-doves on threshed pea stack in June; Suffolk

various plumages resting at the flood tide in September, miles from any other suitable refuge. As neither of these species likes staying in the water very long and both need to rest where they can feel reasonably secure from disturbance, their local numbers would probably have been much less had the refuge not been there. Swallows have nested on this same structure.

While the claims and influence of defence have vastly increased recently, those of sport have diminished. There is, however, a very large area of Great Britain over which sport is still the sole or at least the dominant human interest, and on which other considerations rank second to the preservation of an adequate stock of certain mammals, birds or fishes which can be suitably pursued, ambushed or lured at the right seasons. The broad effect is to confer special protection, subject to controlled destruction, on the game species themselves, to create a specially persecuted class of other creatures rightly or wrongly ranked as "vermin", and to give the most effective protection of all to other species which are graded as neutral and which flourish in the particular conditions arranged for the primary benefit of the game in the area. In this way for example a species like the woodpigeon, although regarded as a pest to agriculture, is assured of the undisturbed use for breeding of a great extent of suitable woodland owing to the fact that pheasants in the same woodlands happen to need strict protection in the interests of sport. The same is true of rabbits, and of course of many harmless forms such as warblers and thrushes.

Some birds, such as woodpigeons, sometimes take food, such as grain, intended to be eaten by game-birds, while others, such as magpies at the pheasant-coops and herons or kingfishers at the fish hatcheries, seek at times to batten on the game themselves at points where human intervention has made them temptingly abundant in a vulnerable phase. But by far the most important effect on bird-life of the game interest has lain in the remarkably effective and widespread system of policing which it has maintained for a century or so against human disturbance or natural destruction of game, and indirectly of those other forms of life which can exist with game without getting themselves blacklisted and suppressed. British gamekeepers form an effective biological police force, employed by private interests to further some purposes which are excellent as well as others which are questionable. Any weakening of this force brings quick and important effects which need to be closely watched.

Roughly parallel to the game-preserving interest is the industry, dominating an even larger area, of rearing livestock. Field and moorland livestock, consisting mostly of sheep, cattle, and horses, assist by grazing and manuring to maintain large areas under certain types and lengths of herbage favourable to a number of species of birds which are thus indirectly dependent upon livestock. A closer dependence is shown by a few species, especially the jackdaw, starling and yellow wagtail, which actually follow the grazing beasts about to gather food and nesting material, and the swallow, which not only catches flies round grazing and resting cattle but strongly prefers to nest in or near cowhouses, stables, and places where livestock are kept. Some birds, particularly ravens, crows and the larger gulls, get fairly frequent chances to act as scavengers of dead livestock, and there have lately been disturbingly frequent reports of carrion-crows pecking out the eyes of living sheep and lambs. Apart from this we have no species which attacks healthy livestock in Britain as the kea parrot occasionally does in New Zealand, except for the raids made by sparrow-hawks and sometimes by other birds on chickens. I vividly remember seeing, at the age of less than five, my first sparrow-hawk swoop down suddenly and carry off my special favourite canary-yellow downy chick out of a few-days-old brood—certainly my earliest bird memory and still one of the most poignant.

Some birds become nuisances by coming down when the poultry are fed and robbing them of their meal; I have seen jackdaws and herring-gulls doing this in Cornwall and rooks are also given to it.

The next great human activity which largely affects bird life is the cultivation of the land. In the long run drainage, clearance of woodland, scrub, bracken and coarse growth, reseeding of pastures, planting or destruction of shade-trees, windbreaks and hedges, provision of accessible water and other development activities decisively influence the numbers and composition of the bird population over very extensive areas. In the day-to-day and season-by-season work of farms and horticulture such operations as ploughing, harrowing, hoeing, sowing, fertilising, spraying, harvesting and fruit-gathering help to determine which birds the land will support and how many of them. The same applies to the cultivation of trees as a crop by forestry, but for convenience that is excluded from the scope of this volume, and is to be handled in the companion book on woodland birds. Disturbance of the soil by various kinds of digging and ploughing and drilling enables a number of birds, ranging from rooks and gulls and starlings to the back-garden robin, to get at insects and other food which they could not otherwise reach. This unintended human aid is the more important because so much of it comes in the winter months when the shortage of accessible food is greatest. Some birds such as rooks and wood-pigeons in the fields and sparrows in the garden give trouble by taking advantage of the easily pierced tilth to gobble up newly-sown seeds or by pecking the tops off sprouting plants. Sparrows in towns make themselves even more unpopular by an apparently insane craze for pecking the flowers off crocuses, polyanthus and other plants designed to look cheerful in spring. This habit is quite distinct from the bud-eating activities at the same season of bullfinches, certain titmice and woodpigeons, which, whatever their results, represent a quite straight-forward effort to get food, including in some cases insects hidden in the buds.

Cultivation brings with it a harvest not only of crops but of weeds, and the birds which concentrate on these, such as the linnet and goldfinch, benefit from and are accepted as beneficial to agriculture. Finches and buntings generally gain much advantage in autumn quite harmlessly by gleaning the cornfields after the harvest is in. Rooks, woodpigeons and sometimes other birds overstep the line by descending upon stooks in the fields and helping themselves to a share of the harvest. On the other side of the account the farmer's methods in cutting crops both of hay and corn have become with advancing mechanisation seriously destructive to some harmless or useful species which nest in the fields, particularly the now local corncrake and quail.

A more serious problem in some places is the destruction by birds of fruit, especially soft fruit. Starlings, blackbirds and throstles probably do most damage in this way on account of their size and numbers and of the fact that the fruit ripens just as their young broods have grown to an age when they are able to rally to the spot in strength and to gobble up cherries, currants, strawberries, gooseberries and raspberries unaided. Later in the year figs and grapes and to a less extent plums, pears and apples suffer in the same way. Many birds take part in the orgy, including jays, magpies and jackdaws, whose spacious crops make them particularly to be feared. Much remains to be discovered about the preferences and food-habits involved; in some areas certain fruits seem to be fairly immune and some believe that the damage is less where birds have easy access to drinking water. Among wild fruits one curious and conspicuous feature is the strong preference shown by many birds for elderberries and the almost universal avoidance of blackberries. This is a direct reversal of the popular taste among human beings.

In the case of fruit grown under glass it is usually difficult for birds to get in where the structure is in proper repair, but warmed glass-houses are attractive cover to small insect-eating birds and there are usually robins to be found, for example, singing and going about their everyday business in the Palm House at Kew Gardens. The problems created for the Great Exhibition of 1851 by the house-sparrows which colonised the original Crystal Palace and covered the exhibits with droppings are a matter of legend, and it is said to have required the tactical genius of the Duke of Wellington to find a solution by having a sparrow-hawk let loose in the building before the opening day.

Farm and garden crops, shrubs and trees form attractive cover for many kinds of birds, and it is in certain types of gardens and orchards that the highest densities of land-bird population are usually reached. Later chapters will discuss more fully the British birds most characteristic of the man-made land environment of farms, gardens, orchards and hedgerows, towns and buildings. This chapter has sought to show in outline how widely and deeply the affairs of birds and men in Britain have become mixed up together. Inevitably the picture has been sketchy, not merely on account of space limits, but because for some odd reason the subject has not so far appealed to many bird-watchers. Some of the most obvious and repaying lines of study have scarcely begun to be followed up, although the answers do not need to be sought by difficult expeditions over the hills and far away but lie almost literally on our own doorsteps. It is important that bird-watchers should look more closely into these matters, not only on account of their scientific significance and their bearing upon ornithological history, but because a better understanding of them would point the way to action for furthering the protection of birds and adding to the pleasures of bird-watching.

CHAPTER 3

FARM LANDS AND FARM BIRDS

FARMS are places where people grow crops, either to be eaten as they are by man or beast or to be converted into some food or raw material by machinery. Farms accordingly consist mainly of open ground, sometimes under herbage which is kept down by grazing, and sometimes under crops such as wheat, potatoes or grass, which are cut or lifted after growing. The farmer needs to maintain and improve the fertility of the farm, its water-supply, and its shelter, and generally also to ensure a certain variety in its pattern. In essence, therefore, the farmer is a preserver and improver of valuable bird-habitats.

The question, however, whether a farm will or will not be rich in bird life depends on many considerations. Some farms, especially in highland or upland areas and in the marshes, are still not very far from nature, being largely devoted to the grazing of types of rough or lush herbage in which the plants natural to the locality are still prominent, while unreclaimed patches and fringes survive freely. On such farms the appropriate waders such as curlews, lapwings or redshanks will often flourish, with stone-curlews at the extreme of dryness and snipe at the extreme of moisture. Skylarks and meadow-pipits from adjoining moorland or heath, tree-pipits from woodland fringes, yellow wagtails from water-meadows, and many other species such as linnets, stonechats and whinchats may be found on such fringe farms, which occupy a very large area in nearly all parts of the British Isles except midland and south-east England, and are in many cases classified agriculturally as marginal or submarginal.

Here the farmer is mostly scratching a living from the land by scratching its surface with the aid of a sparse stock of sheep, cattle and poultry. The main modifications which farming has brought in here are the farm buildings and access roads, some walls, banks, fences and sometimes hedges, some drainage and possibly ponds and tanks, and probably some trees and shrubs. There is some little disturbance to birds from people and livestock and from farm operations, but it is not serious and is probably more than made up by the favourable effects. The effort of adaptation needed from the birds is moderate, but strictly moorland species such as red grouse and ring-ousels, strictly marsh birds such as bearded tits, bitterns and water-rails, and strictly open country birds intolerant of disturbance such as the great bustard cannot normally stand even this degree of interference. The bustard has actually become extinct in Britain, although no doubt more through killing for food or sport than directly through farming. Among species which clearly prefer unfarmed land but can stand a fair amount of agricultural disturbance provided their basic habitats are not eliminated are the stone-curlew, woodlark, meadow-pipit, and curlew. The corncrake adapted itself successfully to nineteenth-century farming methods, but many observers fear that it is finding those of the twentieth century intolerable wherever they are introduced, and it may even in time cease to be a bird of agricultural land. The lapwing shows signs of getting into difficulties also. On the other hand the redshank and the oyster-catcher have lately shown an increasing tendency to spread over farm-land in certain regions.

At the other extreme is a growing area where modern agricultural tendencies are rapidly destroying past links with natural conditions. Pastures are ploughed and reseeded with arbitrarily chosen grasses and clovers. Artificial fertilisers are applied in ever heavier amounts. Horses disappear, to be replaced by tractors and probably to be eaten in the meat-starved towns. Sheep also dwindle, but dairy cattle, pigs, and poultry multiply in new buildings so sterilised and hygienic that they offer hardly a cranny to place a nest or even a fly to be snapped up. Instead of being allowed to grow through the breeding season and then left lying or standing in hayfields, grass is clipped whenever it has grown a little longer than a lawn and at once baked in grass-driers or buried for silage. Combine harvesters, moving rapidly over the corn-fields, take the standing crop in their stride and convert it on the spot into bagged grain. New poisons, which have just been proved to kill some pest not yet immune to them, are broadcast over the earth before anyone has time to test what and who else they are going to kill into the bargain. Shelter belts and windbreaks are clear-felled for timber or to bring in more land, and hedges or banks grubbed up to allow bigger and better ironmongery to move over the ground more quickly and more often. Although many mistakes may be made on the way it is incontestable that many of these developments represent agricultural progress and as such must be supported and endured, yet there is no gainsaying the threat which they represent to many of the birds of farmland, some of which can scarcely be expected to escape a serious reduction in numbers. On the other hand the smaller, busier and more educated modern farm population may well have less time and inclination wilfully to destroy or catch wild birds or take their eggs, and may also, through growing powers of control over rats and other pests, give welcome relief to bird life.

The vulnerability of birds to most of these modern developments depends largely upon their nesting sites. Some farm birds, like the rook, mistle-thrush and little owl nest usually in trees, or, like the jackdaw, starling and barn-owl, in trees or buildings. None of these birds appear to be threatened except the barn-owl, which may easily be poisoned through picking up poisoned rats and mice, while the swallow may suffer through favouring a less hygienic form of byre or pigsty than is now regarded as acceptable by us. Many other farm-nesters are hedge birds, and stand to suffer from a tendency to cut down the mileage of hedges in relation to area, which is very important for density of bird population. On an Oxfordshire farm with several unusually big fields there were 2½ miles of hedge to each 100 acres and census work showed that it was length of hedge rather than area which determined the number of hedgerow birds. Drastic trimming or laying of hedges during the breeding season, road improvements and traffic deaths are all tending to aggravate the troubles of our hedgerow birds, which include such important songsters as the blackbird, throstle, robin, wren and dunnock, while on farms the yellowhammer and by proxy the cuckoo are also largely dependent on hedges to maintain their strength.

By far the most serious plight, however, is that of the ground-nesters. Those which usually nest under some sort of cover, like the partridge and red-legged partridge and tree-pipit, or which breed late like the quail, may not be seriously threatened, especially if they are preserved as game, but other species nesting out in grass or crops, such as the skylark, yellow wagtail, and lapwing may well be eliminated altogether from intensively farmed areas unless they can find sufficient suitable fields undisturbed in April and May.

There are few important farm-birds which are not more or less social outside the nesting season, even to the extent of mingling with other species, and it is possible to meet in one flock any combination of rooks, jackdaws, starlings, lapwings, black-headed, mew and herring-gulls and woodpigeons in numbers ranging up to several hundreds. This sociability, and the mobility which goes with it, is a substantial asset in surmounting difficulties caused by agricultural changes, and one of the unfortunate results of such changes is that they set an even higher premium on adaptability and thus tend to make dominant species more dominant while penalising more specialised forms. Fortunately skylarks, yellow wagtails and lapwings are all sociable and mobile except when breeding, and are likely to be tenacious and ready to seize any opportunity. Yellow wagtails at once bred in Regent's Park when it was closed during the war and lapwings returned in 1943 to breed as soon as cultivation was resumed in Richmond Park, after over a century's absence. The finches and buntings also flock and shift where the going is good.

The birds of our farms are remarkably little changed round the year except for a few summer migrants (chiefly warblers and the swallow, martin, swift, cuckoo, corncrake and quail) and for the wintering fieldfares and redwings. Locally however certain species such as the gulls, golden plover, grey geese, and meadow-pipit visit the farms only in winter, while there may be large increases of others, notably the starling and lapwing. Then again there are birds such as the greenfinch and chaffinch which shift their quarters from woodland to farm, or from one part of the farm to another at certain times of the year.

Having looked generally at the farm environment and the advantages and difficulties with which it faces birds we are now in a position to review the main species which occur on our farms, and the nature of their link with the farmer. Those species which are most important from this standpoint will be more fully discussed individually in a later chapter.

The crows are notable farm-birds, the jay being the only member of the group in the British Isles which has no more than a casual connection with agriculture. The raven in Great Britain is practically confined to the west and north where it is found mainly in hilly and

Robert Atkinson

Newly-hatched stone-curlews squatting in nest in wheat-field on Hampshire Downs. May

PLATE 8

Eric Hoskins

In a Suffolk rookery. April

coastal districts; in these it often visits farms, and according to Ryves (who estimates about fifty pairs in Cornwall alone) cliff-nesting ravens invariably fly inland for their food. While they are fond of devouring the placentas of ewes in the lambing season, and will, like other crows, take any carrion they find about, they do not normally attack living lambs or sheep, and they deserve protection as fine and generally harmless birds.

The carrion-crow, and the hooded crow which replaces it in most of Scotland and in Ireland and the Isle of Man, are commonest on coasts and estuaries, marshes, and hilly areas where there is some cover either in trees or on rock faces. There is reason to believe that both these forms are increasing and locally extending to fresh areas, and there have been disturbing and reliable reports of a growing tendency to attack the eyes of living lambs and sheep in various parts of England and Wales—an unpleasant habit possibly stimulated by the blizzards of early 1947, when millions of sheep and lambs were enfeebled and many succumbed while thick snow-cover made much of the normal food unobtainable. Either crow is a thoroughly unpleasant neighbour wherever there are eggs or helpless mammals or birds about, and on farms they are particularly unwelcome to the shepherd and the poultry-keeper. Arable and mixed farms in the lowlands are often more or less free of carrion-crows, unless they adjoin some more favoured crow harbourage. Then the carrion-crow may outnumber any other member of its family on the neighbouring farms.

The rook is perhaps more thoroughly a farm-bird than any other we have. It penetrates little distance into uncultivated tracts, quits large towns when their suburbs grow into a barrier between farm and rookery, and spends most of its time on farm-land. To a less extent the jackdaw is also linked to farms.

The magpie is less prone to bury itself in thick cover than the jay, but much more so than the field-haunting rook and jackdaw. Thick plantations and hillside woodlands are favourite magpie strongholds, although in many parts of England, Wales and Ireland it has become so plentiful that it is generally distributed even where cover is sparse. On the Continent however it is often both more plentiful and much less shy, not hesitating to nest close to human dwellings, and Irish magpies also are markedly tamer than English ones. In Scotland magpies are very patchy and local. If their recent increase and spread continues they may soon become predominantly farm-birds, but it seems reasonable for the present to treat them as woodland birds which invade the farm freely and in growing numbers, especially in hilly districts and where game is not strictly preserved.

Of birds which feed in the open fields the starling is easily the most numerous, and in winter is liable, in many parts of the country, to outnumber all others put together. It is, however, less completely a farm-bird than the rook, being common in a variety of other habitats also, and there are several prominent farm habitats such as tall crops of all kinds which are little favoured by starlings, their preference being for flattish bare soil or for short herbage. While the fields can never be too small for starlings or the adjoining cover too thick they differ from most other passerine field birds in not caring how far from cover they go. In addition to winning a high proportion of their total food-supply from farms starlings rely on farm buildings and on farm timber for a high proportion of their nesting sites in Britain.

Greenfinches have a strong liking for cover higher and thicker than the ordinary low hedge, but less tall and nearer the open than most of our woodlands. They are not therefore typical breeding birds of farms except where the hedges are really high and thick or where there are orchards or clumps of elders, yews, cypresses, or other attractive tall bushes or low trees. In winter they flock with other finches and buntings in stackyards and stubbles, but still show a distinct tendency to cling to the neighbourhood of cover, to which they retreat on disturbance. They are therefore more typical of mixed flocks containing yellowhammers, chaffinches and sparrows, which share this preference, than of the more open-loving flocks of linnets and goldfinches, although the two groups overlap. They are tempted to the ground chiefly by seeds and fallen berries, and do not therefore show much liking for open grasslands.

Goldfinches take much of their food from thistles and other tall weeds, not minding what the surroundings are like provided such plants are available. They therefore find less sustenance on good farms than on those whose occupiers have been classified by their County Agricultural Executive as B. farmers, or those which the agricultural economist would grade as marginal or submarginal. The more goldfinches you meet on a farm the worse the land or the worse the farmer, or both. In the breeding season, however, goldfinches shift to orchards and gardens or open areas with plenty of trees, and they are then less dependent on large patches of tall weeds.

Linnets are more ground-feeders than goldfinches and are especially fond of stubble-fields, where they gather in flocks of fifty to a hundred, or often more, and mix readily with other finches and buntings. When disturbed they are less given to retiring to the nearest cover, and more liable to rise high and circle round in a close flock, perhaps coming down again a good distance away. Like greenfinches, linnets tend to nest in clusters of several pairs close together, but they nest nearer the ground than any British finch except the mountain-loving twite and are therefore characteristic of coast and upland farms with a minimum of cover, although they occur in suitable hedges and clumps of bushes on many different types of farm, and also on commons and other uncultivated areas. Twites are found on farmland only very locally, chiefly in parts of the north and west of Britain, and normally outside the breeding season.

Chaffinches are woodland birds which have followed the taller hedgerows and amenity plantings out onto the farm and are also often found in farm gardens and orchards. Outside the breeding season they form flocks of one or both sexes, often mixed with other finches and buntings which haunt stubbles and stackyards, usually within easy range of cover. It is no doubt largely owing to the ample winter food which the chaffinch finds so readily on the farms that it is able to rank as in all probability the most numerous British breeding bird. The closely related brambling, which occurs only in winter and is uncommon in many parts of the country, often mingles with chaffinches in the fields and stackyards.

Buntings, like finches, are fond of cultivated land and farmyards in winter, and except for the reed-bunting and the snow-bunting they like breeding on farms also. The corn-bunting is very much a farm bird, and is equally at home on the well-managed highly mechanised large farm of Cambridgeshire or the East Riding and the wretched peasant-holding scraped out among the gorse and heather on a cape in the West of Ireland. This is the more surprising because the corn-bunting avoids nearly all other habitats than farmland, and even on farms is extraordinarily given to concentrating in certain areas and almost ostentatiously ignoring others.

But for the attention it has attracted to itself as a town bird the house-sparrow would be regarded as one of the most specialised of farm colonists, breeding in a very large number of farmhouses and farm buildings in most parts of the British Isles, and foraging particularly in the stubbles, fallows, and stackyards, and along farm roads. Sometimes, especially in company with flocks of finches and buntings, house-sparrows may venture three or four fields away from the farm-house, and this is the nearest they normally get to wild country. The often repeated story that town sparrows move out into the country at harvest time seems to be entirely unsubstantiated. Until recently house-sparrows obtained much grain from horse-droppings and from spilt forage; the replacement of horses by mechanical horsepower has no doubt contributed towards their diminution in modern times on both sides of the Atlantic.

Tree-sparrows, which nest chiefly in holes in trees, are less closely associated with farm-land and much less closely linked with man, although they like to visit stackyards and fields near their haunts.

Skylarks are highly characteristic farm-birds in most parts of Britain, not primarily because they like farms but because they require wide expanses of open country with short herbage, and these are provided by farming more freely than on any other habitat except heaths and moorlands, on which skylarks are also abundant. As sky-larks do not mind nesting in tall crops and are rather late breeders, they may suffer less from disturbance by mechanised agriculture than earlier ground-nesters such as the lapwing. Modern trends in agri-culture, however, do not appear favourable to skylarks and their decrease on farmlands in future years would not be surprising, although few changes in our bird life could be more generally unwelcome.

Woodlarks, even more than goldfinches, are indicators of poor land or poor farming or both, and they also with their mellow notes call attention to poor forestry, since the derelict sites of felled wood-lands are among their favoured habitats. There should however always be a place for woodlarks on steep escarpments and on the fringe of farming country, as well as on uncultivated habitats.

Tree-pipits also find a place on the fringe between farm and woodland, and on hill farms where they go up to or beyond the limit of cultivation; in Wales I have found them up to 1300 ft. and in Shropshire up to 1500 ft. Meadow-pipits, except on passage, keep to rank and coarse types of vegetation, characteristic of moorland and heath rather than farmland, but they breed not uncommonly on farms in damp or hilly areas. Rock-pipits occur on farmland only in certain coastal areas, and then only in fields immediately adjoining the shore.

PLATE III.

V. L. Breeze

Great and Blue tits opening a milk bottle

PLATE IV.

John Markham

Kestrel (hen) at nest in a church tower

Of the wagtails the yellow is much attached to farms in some areas, although it is perhaps more characteristic of marshes and in some parts of the country of heaths and open uplands The future of yellow wagtails on grasslands under a grass conservation policy looks precarious, since they will find difficulty in rearing a brood successfully in fields where the grass has to be cut every time it grows four or five inches long. Fortunately, however, the yellow wagtail is fond of localities which may often prove too marshy, too hilly, too weedy or otherwise unsuitable for such agricultural developments; it also flourishes in legumes and fodder crops, which are more immune from disturbance in spring and may well be grown more in future.

Grey wagtails are little seen on farms except round the streams, ponds, and damp farmyards, but pied wagtails are among the best-known of our farm-birds, and share with starlings, house-sparrows and swallows the habit of nesting in the farm buildings and finding their food on the farm itself. The tendency to eliminate flies and mud from the farmyard and even to have less dung about can hardly benefit the pied wagtail but there is no reason to doubt that it will continue to enliven and ornament many of our farms in years to come.

Tree-creepers, nuthatches and tits, although often seen about farms, have no real claim to be regarded as farm-birds; nor have the red-backed shrike, the flycatchers or the goldcrest, or any of the warblers except perhaps the whitethroat, in so far as it is typical of hedges and hedges are typical of farms. The thrushes often feed on farms, especially the wintering fieldfares and redwings and our resident mistle-thrushes, while blackbirds and throstles are in many parts of the British Isles as typical of farms as of any of their other favourite habitats.

On the Continent whinchats and stonechats will breed in highly cultivated country, but in Britain both, like the wheatear, are more typically birds of the fringe of cultivation. The status on farms of the robin, dunnock, and wren is comparable to that of the blackbird and throstle. Swallows have a more particular claim to be regarded as farm-birds since they so often breed in farm buildings, and are particularly attached to livestock, while as an urban species the swallow has already suffered such a decline that it may soon cease to be a bird familiar to most men, women, and children in Britain. House-martins on the other hand still breed within four miles of Hyde Park Corner and in very many towns and villages, where they are commonly taken for swallows, since it is only in country districts that the two species can be seen together.

Sand-martins, which are confined by their breeding habits to the neighbourhood of sandy banks or cliffs, feed over farmlands within easy radius, but show no preference for them.

Swifts, nightjars, green woodpeckers and cuckoos have a purely casual connection with farmland. Among the owls the barn-owl and little owl both show a marked preference for farms, the barn-owl in particular depending largely on farm buildings for its nest-sites and on the open fields of the farm as a hunting ground.

Among birds of prey the kestrel, and over a growing area of England and in Wales and parts of Scotland the buzzard also, must be counted among farm birds, although both also range over much uncultivated country. Farm buildings are often used as nest-sites by kestrels. In most parts of the country the density of day-hunting birds of prey over farmland is exceptionally and regrettably low, while such destructive pests as rats are excessively abundant, and there are almost certainly too many crows.

Herons and wintering geese and wigeon will feed on farmland, and some of the grey geese are surprisingly dependent on crops as well as grass for their winter keep. On certain islands off the coasts storm-petrels and Manx shearwaters breed in holes in walls or underground on farms, but except where walls increase the number of breeding-sites it cannot be said that there is any significance in the connection.

Three of our British pigeons are more or less dependent on farms for their food-supply, although the woodpigeon and the turtle-dove breed mainly in woods and the stock-dove in any suitable holes on or off the farms. All three are mainly vegetable-eaters and take large quantities of cereals, beans, peas and other crops grown for human or livestock consumption, with a small proportion of animal matter and a varying amount of weeds. Rock-doves also draw some of their food from farms near the sea-cliffs on which they breed.

While there are not enough stock-doves and turtle-doves to constitute more than at worst a minor and temporary local nuisance, woodpigeons are sufficiently numerous to be distinctly irritating and locally damaging to the farmer at certain times, and the Forestry Commission's programme of planting millions of acres of dense young plantations threatens to provide sufficient ideal nurseries for a vast further increase in our home-bred woodpigeon population during the coming decades. The woodpigeon situation will therefore need watching, and there can be no harm in shooting as many of these wary birds as come within gunshot, but it would be rash to conclude that farmers would be ruined or that home production of food would be significantly reduced even should a large increase of woodpigeons occur, as it probably will. Woodpigeons in winter form some of the largest flocks assembled by any of our British birds, and their concentration on certain fields is spectacular; turtle-doves, being summer migrants, build up only comparatively small flocks of a few dozen before their departure in September.

The lapwing is the only wader which has adapted itself to living on farmland all over Britain, although in certain areas of Scotland the oyster-catcher has also done so, and in many inland valleys the redshank has become established during this century. The stone-curlew (see Pl. 7, p. 28) tolerates a fair amount of ploughing in otherwise suitable dry areas, but its eggs are very vulnerable to being crushed or broken, or found and taken by some person or crow as a result of farm operations, and it is therefore important that as many as possible of its remaining haunts should be let alone or grazed rather than tilled. Better drainage of wet lands seems already to have played some part in reducing our breeding stock of snipe, and I was lately told by a prominent farmer of his misgivings after converting a first-rate snipe-shoot into merely indifferent grazing. Curlews also will be affected to some extent by improved drainage of hill farms, but the curlew (see Pl. 3, v, pps. 12, 36) is not sufficiently a farm-bird to suffer seriously on this account. Very locally in a few parts of East Anglia and of Scotland ringed plovers also breed on arable land, both near the sea and far inland.

In his play *Summer Day's Dream*, set in 1975, J. B. Priestley imagines an England whose towns have all been devastated by a third war and which has been recolonised by the great bustard as a bird of downland farms. The bustard survived as a breeding species in Yorkshire and East Anglia some years after the avocet ceased to breed in the nineteenth century, and as it breeds chiefly in crops in Spain and central Europe there is perhaps no insuperable obstacle to its re-establishment as a British farm bird, although disturbance would probably be too great and its occurrences would have to become much more frequent before there would be any hope of this being achieved without deliberate re-introduction and special protection. While such an event is highly improbable it would be rash to treat it as impossible, after the recent invasions of avocets, little ringed plovers and even black-winged stilts.

None of our terns are given to hunting over farmland, as the gull-billed tern does in the Camargue, but several of our gulls, though not the great blackback and the kittiwake, show an increasing liking for farms. Black-headed gulls are found on cultivated land all the year round, having a number of inland breeding colonies from which they range over the surrounding fields like rooks. Many which nest on the Lymington Marshes regularly cross the Solent to forage on farms in the Isle of Wight, while even the market gardens of West Middlesex are now regularly patrolled by birds from the new gullery established about 1941 at Perry Oaks on the edge of London Airport. Among other inland counties having gulleries are Cambridgeshire, Northamptonhire, Derbyshire, Radnorshire and the West Riding. No birds are keener on following the plough than black-headed gulls, and the fast-moving tractor seems to suit them even better than the more plodding horse. (see Pl. x, p. 53).

Herring-gulls also are birds of farmland in many areas, especially in Devon and Cornwall, where flocks of fifty or more birds are frequently seen in the fields some way inland and single birds are passing over constantly even in the breeding season. This is a daytime movement, and before dusk they flight back to the coast. In the autumn of 1932 when I surveyed the Start Point herring-gull roost it was used by about 1400 birds, of which about 200 came from inland on the Start peninsula and a further considerable number from inland behind Slapton Ley. Besides feeding inland many of these birds spend long periods resting in the fields even in the breeding season, and they are also fond of bathing at leisure in fresh water, for example in the River Dart right up to the fringe of Dartmoor near Buckfastleigh. In North Cornwall similar flocks which spent the day in fields in the Wadebridge area flighted to roost on isolated stacks in the sea off Pentire Point.

In other parts of England the herring-gull sticks more closely to the coast and is replaced inland by the rarer species often confusingly called the common gull, but which I shall here call the mew gull, following the practice of Old English, the Shorter Oxford Dictionary and W. B. Alexander. In winter many fields on and under the South Downs between Arundel and Petersfield and the southern Cotswolds

Ronald Thompson

Curlews on a farm

PLATE VI.

C. M. Clark

Oystercatcher nesting in a cornfield

about Tetbury are the hunting grounds of flocks of mew gulls, which spend all day on the farms but return to tidal waters to roost. The lesser blackback also occurs on farms, especially near its inland breeding places and on passage. I have once seen great blackbacks on a farm some way from water in Suffolk.

Neither skuas nor auks appear on farms except by accident or where some attempt at farming is made even in the wild coastal areas where they breed. Of the rails the corncrake is the only species which has made a serious attempt to come to terms with the farmer, and that has had a sad ending. In a Cornish farmyard however I have found some moorhens which had learnt to become almost as tame as in London by associating closely with the domestic geese and poultry around the duckpond, and others have noticed the same habit not infrequently in other parts of the country.

Three of our game-birds—the partridge, the red-legged partridge and the scarce and irregular quail—are almost entirely dependent on agriculture, and all these are described and their relations with agriculture discussed on pages 87-99. The pheasant, a woodland bird, also likes to forage on the farm, and would do so far more if not artificially fed with grain in the woods. The pheasant, however, would probably disappear from many places without elaborately organised game-preservation, whereas the two partridges and the quail can stand on their own feet.

CHAPTER 4

LIFE-HISTORIES OF FARM BIRDS

In the last chapter farmlands were discussed in relation to their bird life, and an outline was attempted of the main species which are important as farm-birds in Britain. In this chapter fuller life-histories are given of thirteen of these species which appear of outstanding significance in considering the bird life of our farms. These are the

Rook	*Corvus frugilegus* L.
Cornbunting	*Emberiza calandra* L.
Skylark	*Alauda arvensis* L.
Yellow Wagtail	*Motacilla flava flavissima* (Blyth)
Fieldfare	*Turdus pilaris* L.
Redwing	*Turdus musicus* L.
Little Owl	*Athene noctua* (Scop.)
Barn-Owl	*Tyto alba* (Scop.)
Lapwing	*Vanellus vanellus* L.
Corncrake	*Crex crex* L.
Common Partridge	*Perdix perdix* L.
Red-Legged Partridge	*Alectoris rufa* L.
Quail	*Coturnix coturnix* L.

Some other species which live commonly on farms but nest in hedges or gardens, or on buildings, are dealt with in subsequent chapters.

ROOK

(Pl. 8. p. 29)

The total weight of the living burden of rooks which our land has to maintain undoubtedly exceeds by a long way the massed weight of any other species of bird. The rook is easily the most numerous large bird distributed throughout the British Isles, and is also one of the heaviest of our common birds, weighing about 16-20 ozs (c.450-570 grammes) the average woodpigeon ranging between about the same limits although a 34 oz. (nearly 1000 g.) woodpigeon is on record. The great bulk of the food of our rooks is taken off the farms. The contribution of playing-fields, commons, railway yards, seashores and other places where rooks feed is relatively small. In fact the distribution of rooks in the British Isles is closely parallel to the distribution of farmers, except that the rook is absent from certain treeless or otherwise unfavourable areas where there is some cultivation. The rook, although feeding and largely living on the ground, rarely brings itself to roost and never to nest upon it, but remains invisibly tethered to the crown of some tree. Yet the tree-based rook is almost as wary of the smaller enclosed spaces and almost as disinclined to come down among tall herbage or standing crops as the field-nesting lapwing. Wide fields, of good soil but with not too much vegetation covering the earth, are its favourite hunting-grounds.

There seem to be in Britain no remaining unsophisticated rooks still living on their own away from man like the starlings of the remote uninhabited islands, or some of the cliff-breeding jackdaws. Starlings and jackdaws, as hole-nesters, find no difficulty in using holes in trees just as well as holes in rock or in buildings, whereas rooks, with their persistent preference for making open nests in colonies, have never been able to divorce themselves from trees. After the Napoleonic wars the London rooks experimented with nesting on buildings, using the weather-vanes on each turret of the White Tower in the Tower of London, and the wings of the dragon of the vane of Bow Church among other enterprising sites, but the practice has never taken hold. Rooks therefore continue to limit themselves, like woodpigeons but unlike other common field-birds, to breeding areas commanded by suitable trees. But whereas the woodpigeon often builds low, and looks to density and darkness of cover for protection, the rook builds high and does not much mind how open and conspicuous the site or how many people pass by, relying on the fact that however many see the nest few will face climbing up to it. This disregard for concealment and quiet works out, paradoxically, as a factor in the protection of the species, which wins toleration and even friendly sentiment by making itself so familiar to man, and in this way gains so wide a choice of nesting sites that it is not generally limited, as so many species are, by the difficulty of finding enough of them.

As a result of much census work and food investigation, leading up to the big wartime Rook Investigation undertaken in 1944-46 for the Agricultural Research Council by the British Trust for Ornithology, much information is available about the numbers, distribution and food of rooks in Great Britain. Just over two-thirds of Great Britain by area were covered under the direction of James Fisher, in the 1944-46 census, and some 900,000 nests were recorded. After making allowances for the margin of error inherent in such large-scale work it can be taken as established that the total rook population of Great Britain at this period was probably over 2¾ million but under 3 million, of which between 1¾-2 million were in England, roughly ¾ million in Scotland and roughly 200,000 in Wales and Monmouth. There were accordingly about 50 rooks per 1000 acres on an average over the country as a whole, or just over 30 to the square mile. This average, of course, irons out wide differences between the more and the less favoured areas. Over England as a whole rooks were half as dense again as over Scotland or Wales as a whole, but Scotland and Wales have relatively much larger areas uninhabitable for rooks, and when these are eliminated the picture is very different. In England large regions in the plains and lowlands show densities of over 75 rooks per 1000 acres, but only two small areas, in Staffordshire and Northumberland, exceed 300 per 1000 acres; whereas, in Scotland, every lowland county, all the other east coast counties up to Moray, and also Caithness show densities in this class. This means about one rook to every three acres, a remarkable abundance for so large a bird, seeing that an acre is no more than a square of seventy yards along each side. But the density of the Scottish rooks over the ground is not so remarkable as the density and size of the Scottish rookeries as compared with the English. Although Jesse in 1832 recorded the Hampton Court rookery as having averaged 750 nests for the past four seasons there is no known rookery of this size surviving in England at the present time, and even a colony of 500 nests is exceptional; (for census purposes groups of rooks' nests separated by 100 yards or more are treated as distinct rookeries). Yet in Scotland there are plenty of colonies running up to 1,000 nests or more, and the rookery in the Crow Wood at Hatton Castle near Turiff in Aberdeenshire gave a total of 6,085 nests at the 1945 count. This is probably the largest breeding colony of any land bird in Great Britain; for comparison it may be recalled that all the heronries in England and Wales rolled together would only make about two-thirds of this total, while only the two largest British gannetries exceeded it in 1939. One German rookery (Poggendorf) was claimed to contain roughly 9,000 nests at the end of last century and two others to total 6,000 or more. In no other country, however, have such high densities of rook population been recorded as in eastern and southern Scotland. This is all the more remarkable seeing that Scotland is the most northerly country wholly included within the rook's breeding range, and that in the West Highlands and Islands the rook is a scarce and declining species, except for a recent colonisation at Stornoway. There, in 1939, I found well over 200 nests at the Lews Castle, but was unable to make an accurate count owing to the number of weaver-bird-like masses of communal nests crowded side by side and one above another, so as to fill the greater part of the crown of a tree with a great conglomeration of twigs.

A rook census of Ireland remains to be undertaken, but there is certainly a high density in many parts of the country; on a transect from the train between Dublin and Connemara over twenty years ago I found an average of 6 rooks per mile compared with only 3½ in Midland England and less than 3 in winter in Northern France. East of the Shannon Valley the average was nearly 10 per mile. In other parts of Europe the rook is not nearly so dominant. Even in France it extends as a breeding species only down to Auvergne and the Lyons area. It becomes scarce in south Germany and north Italy, and reaches Switzerland only in winter. It stretches south-eastward across Yugoslavia to Bulgaria and south Russia, but stops short of the Urals. The northern part of the Continental breeding range is deserted in winter, and migrants shift down as far as the Mediterranean.

The rook therefore is a bird confined to quite a narrow belt of the Northern Hemisphere stretching from the Atlantic through some seventy degrees of longitude eastwards, and from the Balkans through about twenty degrees of latitude northwards. It takes some effort to realise that a bird so familiar to us is only of seasonal or rare appearance over so much of Europe, and drops out as a breeding species within four hundred miles southward and within two thousand five hundred miles eastward of our shores. Except in a few parts of France I have never personally found rooks on the Continent in summer in numbers which would be considered normal in England. Whether to be thronged with rooks is a blessing or a curse, it is certainly something which the British Isles enjoy or suffer beyond all experience elsewhere.

This question of the beneficial or harmful character of the rook has been debated and pronounced upon with fervour for many years. While one large museum exhibited the rook among birds "specially beneficial to agriculture", Dr. W. E. Collinge, after examining the stomach contents of 2,000 specimens, pronounced it definitely injurious. In my *Birds in England*, published in 1926, I criticised this practice of jumping to conclusions on insufficient evidence, and in *The Rookeries of the Oxford District*, published in *Journal of Ecology* in February 1930 my brother and I showed, on the basis of a census, that there seemed "no reason to suppose that the depredations of the rook on cereal crops in the Oxford area are anything but a negligible factor" when the maximum amount which the birds could consume was set against the quantity grown. Subsequent research relating the consumption of food by the rook and other species to the supplies locally available has provided a growing weight of evidence to the same effect. Looked at nationally the proportion of food which is prevented from reaching human consumption by the depredations of rooks is negligible, while on the other hand it remains to be proved that insect pests would be more numerous and would do more damage if no rooks existed. And these after all are the material questions.

The only practical economic importance of knowing what proportions rooks eat of say grain and leatherjackets is to guide us to a true judgment whether our grain and other crops might be increased if rooks did not exist or if their numbers were reduced. We are not concerned, either as regards grain or leatherjackets, with fixed quantities from which any subtraction must correspondingly reduce what remains. We are dealing, on the contrary, with the dynamics of reproduction and population-increase in plants and insects of enormous fertility. The capacity of either an ear of wheat or a pair of craneflies to reproduce their kind vastly outruns the capacity of the earth to absorb the annual increase, and much of their fertility is doomed to be wasted, if not by one means, then by another. Even if such a bird as the rook ate nothing but wireworms it would still be of no real benefit to agriculture if there were not enough rooks to destroy enough wireworms to make any difference at the end of the season to the number of wireworms left. On the other hand, if it ate nothing but grain from sowing through harvest the rook would do no significant harm if the amount harvested were nevertheless as much as the land could yield on account of limited sunshine or fertiliser or other factors. In such cases the main economic role of the rook would be to form a partial alternative to other channels for frustrating the excess reproductive capacity of the grain or insects which it consumed.

There is every reason to suppose that this is substantially what occurs, and that those who have laboured to prove the rook the farmer's best friend are probably just as wrong as those who have clamoured to wipe it out as his worst foe. Even if, after many more years of intensive research, it becomes possible to show that given a different rook population a given crop in a given area might have amounted to more or less, we are still for purposes of practical policy no better off, unless we can also show that the result would hold good not only in one particular season or for one crop but over a long run of seasons and a wide range of crops, or that it is possible to forecast the special conditions in which rooks will inflict measurable damage and to guard against it effectively and economically. Had the evidence been otherwise, and the rook proved to be on balance harmful, it by no means follows that any practical means could have been devised of achieving and maintaining a substantial reduction in the numbers of so widespread and mobile a creature without incurring economic burdens much greater than would have been involved in letting it alone.

Much time and effort have been lavished on probing the economic status of the rook, and there seems to be some disappointment that the rook should not as a result have been solemnly pronounced guilty or not guilty. Those who feel such disappointment forget that this is a matter of science, not of justice, and that we are considering not a brief series of acts by some individual but the entire way of living of a species which has become through the centuries closely interwoven not only with the wild life of Britain, but with its agriculture. If anyone has been found guilty in the rook investigation it is the jury—that well-meaning body of men and women who persist in personifying birds and saddling them with human attributes, who seek to tear out some small fragment of the living web of nature and to pass judgment on it out of its context, and who see no distinction between transactions with fixed amounts (such as importing oil or robbing a bank) and the dynamic interplay of animals and plants which have contrived to evolve and to multiply very satisfactorily by eating one another in reasonable quantities as they go along.

It is impossible to summarise briefly the very complicated material now available about the rook's diet, but two or three examples may be given of the dangers of being dogmatic. In the Rook Investigation more than half-a-million rooks were watched feeding, and taking the whole year round more than half were on grass, with a peak of 77 per cent. in June. About a quarter were seen on stubble and corn together, the peak for corn being in April (31 per cent.) and for stubble in September (34 per cent.). Yet the aggregate volume of stomach contents of 1,577 birds also collected round the year from various parts of England and Wales in 1944-45 contained 86 per cent. grain and only 5 per cent. animal matter. But grain is digested far more slowly than animal matter, and after making allowance for this the percentage of grain found drops below 75. These, however, are all adult or fledged rooks, and in estimating the effect of the species a further adjustment may have to be made for the proportion of insect food carried to the nestlings which may be larger, though there is evidence that it is not very much larger. The next stage of analysis is to distinguish how much of the corn consumed by rooks would otherwise have been eaten by man or by his livestock. In September observation shows that six-sevenths of corn eaten by rooks is from stubble, and by far the greatest part of this would have simply gone to waste in other ways. Again, a surprisingly large amount of corn is found and eaten by rooks in spring and summer after the new crop has got well under way. Hardly any of the corn remaining in the open so long as this after the last harvest would go to human consumption and much of it is already spoilt or would be wasted. The fact is that corn-growing in Britain is inevitably a wasteful occupation, in which a good deal more seed must be sown than can possibly grow, more grain must be cut than can be brought in from the fields, and more must be stacked than can be fully threshed, quite apart from such common tragedies as corn lodged by wind and rain so badly that it cannot be properly reaped. These facts are obvious to anyone who goes about the country with his eyes open, and they make it rash to claim that any considerable proportion of the grain which is undoubtedly taken by rooks is really taken out of our own mouths.

In addition to their wide normal range of plant and invertebrate food rooks will sometimes eat small mammals and birds and their eggs,

Eric Hosking

Skylark

C. P. Rose

Carrion-crow at nest

but they do not compare with other crows as destroyers of eggs and young. On a main road in Berwickshire one July morning I disturbed a couple of rooks from the corpse of a rabbit just killed by a car. When I stopped one came back and began tearing off gobbets of flesh amid a cloud of flies. A juvenile rook, after being warded off at first with hawk-like gestures, was then allowed to share, and three more adults joined in later. Suddenly a curlew in a neighbouring field gave an alarm, and all leapt to attention and flew off. I found that the rabbit's eyes had been picked out and several gobbets were lying about the road. The next day in Dumfriesshire I saw two in the middle of a tarmac road feasting on another rabbit, and another on a high moorland road apparently pecking at horse-dung. In Northumberland I have seen as many as 32 rooks and jackdaws feeding avidly on the roadway, again apparently on dry horse-dung. During caterpillar plagues the rooks will gather to feed on them in oaks and other trees.

Among the many birds which flock in our fields the rook is almost alone in maintaining the same close community through the breeding season as at other times of year. The jackdaw also remains sociable throughout the breeding season, but on account of the greater difficulty of finding plenty of nest-holes together, jackdaw colonies tend in many areas to be smaller or more dispersed than rookeries. This continuity must have a powerful influence on the rook community; a rookery is a flock settled down at home, while a flock is equivalent to a rookery on the move. But there is good reason to believe that the rookery is itself only a part of a larger and often longer-established rook community formed of the birds which resort year after year to roost at the same traditional site. W. B. Alexander in his paper on *The Rook Population of the Upper Thames Region* showed that where these roosts are not in or near a large rookery they are often marking the site of a former rookery now abandoned. These roosts may include from 5,000 to 12,000 or more rooks, and not infrequently a similar number of jackdaws. They therefore greatly exceed in size the average rookery, although as roosts they cannot compare in scale with the larger starling assemblies, and they do not draw birds from such long distances.

As there is much coming and going at certain seasons between the rookeries and the roost it is possible to discover which rookeries are the breeding settlements of rooks belonging to each roost, and thus to map the main "tribes" of rooks with their headquarters and their rookeries, although it would not be right to speak of their territories, since apparently rooks from one roost will forage up to the neighbourhood of another. Oxford, for example proved to be partly in a no-man's-land between as many as five different tribes of rooks, with their roosts, strategically placed, commanding the lower slopes of the Cotswolds (near Charlbury on the Evenlode and Kirtlington on the Cherwell), of the Chilterns (Rycote near Thame and Chiselhampton near Dorchester) and the Berkshire Downs (Marcham near Abingdon). At the end of the day the birds assemble in trees near their feeding places and flight to roost at no great height, sometimes stringing out in long columns which may extend the procession over as much as a mile. They fly late and sometimes only get in after dark. Occasionally they roost on the ground; I once flushed a large silent flock well after dusk on a frosty December day on one of the highest points of the Cotswolds at about 950 ft., which when disturbed rose noisily and gave every appearance of having been settled down for the night there.

Although rooks like to feed quite close to the rookery they will cling to a favoured site even though all the land within three miles is sterilised by building. In Inner London rooks went on nesting in the City until a century ago, and the historic Inner Temple rookery was even recolonised without success as lately as 1916, a year after the Gray's Inn colony faded out. In Westminster and Kensington regular breeding ceased about sixty years ago, although sporadic attempts round Hyde Park continued into this century. In 1926 I noticed a still occupied small rookery near Earlsfield by the Portsmouth main line about 5 miles from Hyde Park Corner; going northwards at that time rooks became fairly plentiful from Alexandra Palace on. By 1945 only one rookery survived in the County of London, and there were only two others anywhere within 10 miles of St. Pauls. In 1928, and perhaps later, there was still a small occupied rookery on the Quai du Louvre in the heart of Paris, about three miles from the nearest open country.

In some places, as at Stornoway, rooks will fly across salt water between the rookery and their feeding ground, and in coastal districts they not infrequently take to feeding on the tideline. Normally rookeries are so sited that the feeding-grounds can be directly reached without surmounting obstacles, but the rookery at 150 ft. altitude in the valley of the River Mole at Box Hill in Surrey has a regular flyline over the 600 ft. ridge of the hill, beyond which the birds drop to feed on fields down the slope at 300–400 ft. At Little Knoll, Maiden Bradley on the edge of Salisbury Plain I found a large rookery running up to the summit of the steep hill at 833 ft., although there were plenty of alternative sites lower down. A rather similar example in Dovedale was a rookery of more than two hundred nests at Twelve Apostles in a very steep wood between 700 and 900 ft. above sea-level. Many of the rooks flew over the 1,000 ft. ridge to feed in the Manifold Valley or near the Izaak Walton Hotel 500 ft. or more below, while others foraged on the higher fields at 1,000–1,100 ft., and yet others flew along the face of Thorpe Cloud on the Derbyshire bank. In Shropshire I have seen recently fledged young rooks with their parents in June at 1,250 ft., and in July I have found rooks, together with starlings, common in scattered parties up to nearly 2,000 ft. on the Cheviot.

There are few if any kinds of trees in which rooks will not build, and they nest fairly readily in whichever kinds are commonest in each district, and are high enough to give reasonable protection. I have seen as many as four occupied nests in a dead beech without even any bark remaining on it, and two in another, although there were plenty of living trees unused near by. Bleak exposed sites or places subject to much disturbance are not avoided, and even severe persecution does not readily compel the birds to desert. Rookeries therefore are remarkably stable from year to year, the abandonment of old sites, the colonisation or budding off of new, and the rise and fall in the strength of colonies being usually on a moderate scale.

Rooks are curiously averse to nesting in large woods, even along the edges, and they prefer fairly open groves or windbreaks and hedgerow trees to thick copses or plantations. In some rookeries there may be an average of little more than one nest for each occupied tree, while in others nests are densely packed with thirty or more in a single tree.

In gathering sticks for building a rook will risk a perch hardly sufficient to bear its weight and will crane forward with half-open wings and fanned out tail to snap off twigs, many of which fall to the ground and are given up. I have seen one thus engaged topple right over and flutter about upside down, without letting go its perch. Sometimes the birds go down and retrieve sticks from the ground beneath the nest. Attention to the nest is not confined to the breeding season, and I have seen a nest in the Isle of Wight newly built at the end of October. The eggs, usually from three to five, are variable but show a wealth of ashy-grey and brown markings on a more or less pale greenish ground. They are laid rather early, in late March or April, and are hatched by the hen in from sixteen to eighteen days, after which the young take almost a month before fledging. Parents may be seen feeding young as late as the second half of July, thought there is only one brood. The nests house many small parasitic creatures; one which I sent for entomological analysis from near Oxford proved to contain six species of beetle (one being characteristic of birds' nests), four species of flies, and one each of spiders and three other groups, to a total of nearly sixty individuals.

More serious are the parasites living on the rooks themselves. In the course of the Oxford rook census in 1928 a check was undertaken to account for mortality among young birds, and out of 33 examined 31 or 95 per cent. were infested with gapeworms (apparently *Syngamus trachea* Montagu), as many as forty pairs of worms being counted in a single fledgeling rook. Out of eight adult rooks examined only four were carrying gapeworms and two of these had only a pair each.

The rook is at a distance a totally black bird, showing purplish gloss or iridescence in a strong light; the wings are broad and fairly long, generally showing an array of separate "fingers" in flight; the fairly long tail expands towards the tip; the bill and its bare base, except in young birds, are a whitish, dried bone colour. Individuals vary a good deal in size but the average rook is smaller, shorter-winged, shorter-tailed and less deep in the bill than the average carrion crow. On the ground the rook walks or struts rather more deliberately, but like the crow hops clumsily in haste; the loose flank feathers give a characteristic and distinctive baggy look to the rook at rest. In the air the thick neck is contracted to form one line with the body and bill; the ordinary flight is direct and rather laboured, with continuous wing-beats usually in rather faster time than a crow's. Outside the breeding season, especially in a high wind, rooks will play in the air with remarkable speed and agility and with conspicuous enjoyment. Jackdaws often join in these games. The rook's normal speed of flight seems on evidence so far available to be considerably faster than the carrion crow's, ranging from about 27 to 35 miles an hour, against 20-32 for the crow. The rook has a remarkable voice, capable of a wide variety of more or less discordant noises, with a much greater range of pitch than the carrion-crow's, but without its rhythmic and sonorous qualities. It may well prove that this ample range of the

S. C. Porter

Cock corn-bunting on singing post

Robert Atkinson

Skylark's nest in field of barley; Hampshire Downs

rook's power of expression enables it to communicate more fully and effectively than most other species, but our study of rook language has not yet advanced sufficiently for the subject to be profitably discussed here.

CORN-BUNTING
(Pl. 9, p. 48)

No bird looks more dull and uninteresting at first sight than the heavy, clumsy, streaky-brown corn-bunting, but dull as it may look the corn-bunting is a bird of strange secrets and surprises. Even its distribution is full of mystery. All it appears to need is a stretch of open ground, either on hill or plain, cultivated or rough herbage, or waste, with a few higher perches either on a bush or spray or wall or fence or telegraph, or even a building or tree. Neither exposure or shelter, nor presence or lack of water, nor liability to disturbance by men or vehicles or animals seem to matter much. Yet despite this apparently undiscriminating taste the corn-bunting is one of the most difficult birds to satisfy about a place to live. Dotted up and down the British Isles, from Carloway on the Atlantic coast of the Outer Hebrides or Slea Head, the westernmost point of County Kerry, over to the easternmost strip of East Anglia and the southern coastal plain of Sussex, or along the coasts of the Moray Firth patches of corn-buntings keep cropping out, far away from their nearest kindred. Sometimes in a few fields, sometimes for miles on end, the species is not only present but common; then an invisible line is crossed beyond which there are again no corn-buntings to be seen. On one leisurely summer journey from London to Scotland I kept a close eye on the corn-bunting patchwork, which is easily traced as the birds, wherever they flourish, sing so loud and sit so conspicuously on commanding perches. The first to be met going north were straddling the Bedfordshire-Cambridge-Huntingdon border between Potton and Eltisley and from there on they were pretty well continuous across the Isle of Ely, over the fens into Lincolnshire through Boston and up the coast to about Saltfleet. From there on I saw none along the Humber to Goole, and only one down the Yorkshire side until Hull. They reappeared strongly about Hedon and were very common from there to Spurn Point and on up the Yorkshire coast to Scarborough, as many as seven being seen along a mile of road at one point. Turning inland I found no more south of the Tees, and in Durham only one near Hartlepool, but up the

Northumberland coast they were frequent from near Alnmouth past Bamburgh to Beal opposite Holy Island. It appears that a map of British corn-bunting population would consist of a vast number of distinct patches, some small, some very large, separated by large blank areas, with a tendency for the patches to cling to the coastline, especially where it is well cultivated, and to the richer and more open inland plains, valleys, or lower uplands. Probably the main strength of the species is concentrated along a broad belt from the Wash down through the Fens, thinning as it follows the northern slopes of the Chilterns and crossing the Thames on a seven-mile-broad front from south of Wallingford to north-west of Dorchester. Thence westward it follows the Vale of White Horse tapering towards White Horse Hill where it ascends the slopes to the Ridgeway to reach the Marlborough Downs. There are certainly many outliers of this great corn-bunting belt, and there are some interruptions in it, but an ornithologist crossing it when the birds are in song can hardly fail to be impressed by their sudden abundance between the imaginary lines which bound it, and which deserve to be precisely surveyed. H. A. Course, who made a census of six square miles of corn-bunting country on the borders of Hertfordshire and Cambridgeshire in 1938, found 72 singing males with a density on the arable section averaging about 25 males per 1,000 acres. So far as can be judged from superficial observation the birds cluster in the same areas year after year, although it seems that a decrease is gradually taking place and former haunts are in some cases being abandoned.

A similar patchiness is characteristic of this species abroad. For instance on a journey by the Rhine to Italy corn-bunting country was entered at Emmerich just after leaving Holland and ran from there nearly to Cologne, above which none were seen all through the gorges of the Rhine, but another patch was struck between Heidelberg and Bruchsal in Baden, and then no more right through Bavaria and Tirol to Milan. Similarly in France I have noticed patches of corn-buntings in Picardy and again near the Seine south of Paris and no more until right down at the mouth of the Rhone. Making full allowance for hasty observation there are certainly gaps varying between a few miles and a hundred or more between some of these patches, in which the corn-bunting immediately becomes abundant.

Arbitrary and inexplicable as they seem these patches are too clearly marked and enduring to be accidental, and they represent a challenge to research. Meanwhile a few suggestions may tentatively be put forward. The reason why corn-buntings are not usually found thinly scattered or in isolated pairs, but tend to occur densely or not at all, may well be connected with their extraordinary polygamous habit, discussed later. If mutual stimulation among males is a factor in breeding success, that also would encourage bunching. As almost the entire foraging for the young is left to the hen, and several hens have to rear broods close together, an exceptionally rich food-supply is presumably required to ensure survival of the young in unfavourable weather and this is secured partly by the unusually late breeding season and partly by concentrating in favourable areas. Rich soil, as in the alluvial valleys of large rivers, or on sheltered coastal plains and fenland, is often rich in corn-buntings. I can think of no corn-bunting locality known to me which is not on pretty good farming land except at certain points on the coast, particularly in the milder parts of the country, where closeness to the sea appears for some reason to reconcile the birds to sites which would not seem attractive inland. Certain soils on the chalk and upper greensand are particularly often favoured.

Next to the cirl the corn-bunting is the most southerly in range of our native buntings. It lives in the countries round the Mediterranean, the Black Sea, and the Caspian and northwards to the Baltic, with only small outposts in southern Norway and Sweden. In Britain it continues in some strength up to Shetland where are the northernmost of all its settlements. It does not extend far into Asia, and has a tendency to migrate southwards in winter. In Britain it disappears from many of its haunts at the end of July or in August, and there is a good deal of emigration, as well as immigration from the Continent, although the species is resident in most parts, with only local movements connected with the break-up of territories and the formation of winter flocks. Sometimes these movements are very pronounced. In north Cornwall, in an area under continual observation during the first half of September none were found except on one morning before breakfast, when a party of about ten were busy in a gorsy field quite close to houses; a few hours later they had vanished.

In the breeding season male corn-buntings hold territory, in which nesting takes place, but food is often sought far outside it. The extraordinary habits of this bird were first guessed at in a paper published in *British Birds* in March 1932 by John Walpole-Bond, and were conclusively and fully established during 1932–34 by Lieut.-Col. and

Mrs. B. H. Ryves, whose observations were published in *British Birds* in June and November 1934. They showed that of fifteen cocks watched intensively in 1934 in Cornwall four had two hens each, seven had three, two had four and two had as many as seven, these being minimum figures, giving at least 51 hens to 15 cocks. In one case five hens belonging to a single cock built nests and hatched young almost simultaneously within a radius of about 15 yards. Although the hens were single-brooded about two dozen of the young of this one cock were reared in the season. Territories were not permanently taken up by the cocks until after the middle of May, and were abandoned within three months or so. The pertinacity of the cocks in sitting up on prominent singing stands where they can perpetually advertise their presence and keep unbroken watch on what their mates are doing becomes understandable when their polygamous way of life is known. Although apparently lethargic, not given to helping with nest-building or incubation, and only rarely with feeding the young, the cocks are actually very wide awake and always ready to chase their mates back to their duties at the nest.

The nest, placed in low rank vegetation or shrubs usually a little above the ground, is large and loosely built of grasses and bents and sometimes dead leaves, lined with finer grasses and roots. The eggs, most often from 3 to 5, are laid sometimes in late May but more often in June or even July, there being no other resident species so late in starting to breed in Britain. They are boldly marked with the typical bunting streaks of "writing" on a light greyish or brownish ground, and take 12 or 13 days to hatch and only 9 to 12 more before the young fly. One or at most two broods are usual, but cases of three have been recorded.

Corn-buntings eat seeds and fruits, caterpillars, ants and other small creatures indiscriminately, but it appears that the bulk of the diet is vegetable. Although so fond of perching on overhead wires the corn-bunting is most unskilful at it, constantly swaying and struggling for balance in even the slightest breeze, with the curved toes gripping grimly rather than firmly and the legs flexed back flush under the body, the tail being stifly depressed nearly at right angles to the belly in an effort to compensate for the clumsy stance, which is continually shifted. In flight the wings sometimes appear to be raised almost at right angles above the back and brought down through an arc of nearly 180 degrees. Although so ungainly-looking the flight is quick and

H. G. Wagstaff

Buzzard eating a rabbit

PLATE X.

Eric Hosking

Black-headed gulls following a harrow, Inverness-shire

direct, with few and brief interruptions and undulations and is usually fairly low. In the breeding season it often ends with or includes a long shallow downward glide with outstretched wings, and the cock often forgets or disdains to draw up his legs on taking flight, and looks more ungainly than ever flying along with them hanging straight down.

The song is loud, far-carrying, brief, and to our ears most unmusical, although it has as clear and effective a pattern as the wood-warbler's or the chaffinch's. It opens hesitantly with several *tip* notes (which sometimes contain a suspicion of the ringing quality of a chaffinch's *pink-pink*) and becomes shriller as it gathers speed first into a rapid stutter and then into a confused explosive jangle, which has been compared to the jingling of a bunch of keys or the smashing of glass, and is at least as removed from traditional Western ideas of music. The entire performance is over in two or three seconds, and varies only within narrow limits, but whatever his demerits as a songster may be the corn-bunting is almost unmatched for persistence, churning out his refrain with great regularity between six and eight times a minute; one timed in Berkshire began again like clockwork every seven or eight seconds. It carries well over a quarter of a mile in favourable conditions, and goes on almost all day from about mid-February to August, and occasionally during the rest of the year, though I have never noted full song in September. I have heard corn-buntings singing among snow in the Crimea in February and at Avebury in March. In Germany I have seen cocks singing from trees, once actually within a wood, and I have also observed song delivered from a clod in a field in the Vale of White Horse: from a ruined chapel and a modern metal windmill in Suffolk, and on the wing both in England and in Germany.

In the breeding season the most frequently heard note is a hard *tip* or *chip*, but afterwards this is replaced by an excitable penetrating clipped note, often doubled or tripled, reminiscent of the American cardinal's call, or more distantly of a hawfinch's clicking note.

Compared even with a cock house-sparrow or a greenfinch the cock corn-bunting is a very plump, heavy, large-billed bird, but the hen is appreciably smaller. The head is very flat, its profile almost continuing the line of the upper ridge of the bill. The plumage is ragged-looking and dingy, the upper parts being of varying shades of dull pale brown with a mass of dark and creamy streaks, while the buffish underparts are also deeply striated with brown down the flanks and on the breast, which in adults has a dark V-shaped mark beneath the yellow chin, and in some birds a heavy blackish-brown half-collar. The legs are yellowish flesh-colour: the bill yellowish, stout and flattened, but curved along the upper ridge; in song when it is opened to its widest the more orange gape is displayed. The slenderer, buffer, less boldly-marked juveniles can be puzzling in late summer or early autumn if met with apart from adults, especially as they are apt to skulk.

SKYLARK

(Pl. VII, p. 44)

Skylarks, when they are not on the wing, like to have their feet firmly on the ground, and are remarkable among passerine birds in not only refusing to perch on trees but in usually keeping a good distance away from them. Even to see a skylark under a tree is quite out of the ordinary, and they are almost equally reluctant to approach hedges, or rock-faces, although they will not infrequently perch on fences or overhead wires, and occasionally also on walls or bushes. While broadly speaking skylarks may be described as birds of wide open spaces, and while they almost invariably avoid tall woody cover of all kinds, they are not like lapwings in confining themselves to herbage and crops low enough to allow them to look over the tops when they want to. It is not unusual to find skylarks in vegetation thick and tall enough to block their view of everything except the sky above. Such cover is however rather tolerated than sought, and in a country where vegetation grows so freely skylarks would be severely limited in their distribution if they were unable to breed and forage in herbage or crops a good deal taller than themselves. The steppes and grassy plains of eastern Europe and of Asia offer probably more perfect lark habitats than any in Britain, but skylarks have shown a good capacity for adapting themselves, and are firmly established in open country throughout the British Isles. They seem equally at home in arid and in damp conditions, and are remarkably little troubled by strong winds and exposed sites. Being ground-feeders and ground-nesters they are rather vulnerable to disturbance, and are therefore liable to be missing from open spaces close to large towns unless agriculture or sport has restricted access by people and dogs.

On the rough hill pastures and sheep-moors skylarks breed up to high altitudes, especially where there are grassy patches. I have found them flourishing up to more than 1,800 ft. in both Ireland and Wales, and up to 2,000 ft. on the Cheviots, where I came upon a nest at above 1,600 ft. In such districts the skylark is often a purely upland bird, not breeding in the valley bottoms or on the lower slopes.

Peat-bogs, sand-dunes, tidal salt-marshes and other non-agricultural open tracts are often rich in breeding skylarks, which also inhabit most types of arable farmland and some types of pasture and meadow. Similar artificial habitats such as golf courses and airfields are also promptly colonised. In the short period after it was a wood and before it was submerged under many feet of water the bed of the Queen Mary Reservoir at Littleton was occupied by many skylarks. Unfortunately neither the skylark nor any other species has filled in Britain the ecological niche occupied on the Continent by the crested lark, which has made itself at home in railway shunting yards and sidings, barrack squares, roadside waste patches, building sites and similar artificial open spaces.

Although fairly large as passerines go, skylarks are easily the smallest common birds which actually live the year round getting all their food in the fields on our farmlands. They weigh only about an ounce-and-a-half, about as much as two robins, and about half the weight of a starling. That they are so well-known, in spite of their inconspicuous plumage and small size, is largely due to their intensely territorial way of life, which encourages a vast outpouring every spring and summer of the rich vigorous song, and also leads to those incessant pursuits and skirmishings called skylarking. Being normally without lookout posts or singing stands they have been compelled to make the sky their watch-tower, and to spend on the wing during the season a proportion of the daylight hours which is matched by few if any other species except those taking flying prey. Starting before daybreak and continuing until after dusk the song-flight and the aerial skirmishing give an impression of acute competition for breeding space, in great contrast to the leisurely and ample circuits of the woodlark.

To a less extent the great flocking and migratory movements of larks, and their reputation as food in many parts of Europe have helped to make them familiar. It is only fairly recently that the wholesale killing of skylarks for the table has died out in England. Knox, writing only a century ago, describes how at that time near Brighton as many as twenty shooters might be out after larks in a single field, luring them with a device of revolving mirrors called a "lark-glass" which seemed "to possess a mysterious attraction for the larks, for they descend in great numbers from a considerable height in the air, hover over the spot, and suffer themselves to be shot at repeatedly without attempting to leave the field or to continue their course ...... notwithstanding the crowd, and the noise of voices mingled with the continual roar of guns, the infatuated birds advance stupidly to their doom. . . ." Knox goes on to refer to the heaps of dead and dying larks round these spots, and to the quite distinct and much more skilled operation of netting larks on dark nights between October and March, also round Brighton "where these birds form a very considerable article of traffic and hang in numerous bunches at all the poulterers' stalls in the town and market".

He notes that in fine weather the larks roost in the stubbles, but in wet weather they lie in thick rank meadows and along the higher brows of grassy fields, choosing the exposed side of a hill by preference. Fortunately the migratory flocks of larks no longer have to suffer such slaughter in England, although it still continues in parts of Europe, and no doubt takes toll of our breeding birds, many of which move south for the winter. Despite this exodus our winter skylark population is greatly swelled by vast numbers of immigrants from northern Europe, and also from central Europe, although most of these pass on to other winter quarters.

This skylark migration is more readily observed in daylight than that of most other birds, especially when the flocks pass over areas where skylarks are not usually seen. There was a particularly strong passage over London in October 1925. I first heard the cry overhead among redwings towards midnight on the 16th and the next day found some on the grass near the Round Pond in Kensington Gardens. The westward passage increased throughout the following week, and on the 25th I counted more than three hundred crossing the Round Pond in flocks and straggling parties within forty minutes, and estimated that at the very least 50,000 skylarks must have crossed Kensington Gardens during the previous ten days. From the 26th the movement tapered off, although a few continued to pass up to November. An even larger movement southward passed over on 15 January, 1926, when I counted 315 passing over in 15 minutes in the morning on a hundred-yard front at a height of about 150–250 ft. and later saw over 180 pass in three minutes. This movement went on all day, very silently, and even on the most conservative estimate can hardly have involved less than a quarter of a million skylarks. The weather was severe, with heavy snowfall following three days of intense cold and an east wind. The next day, which was little less rigorous, practically none were to be seen.

Skylarks range right across the open lands of Europe and Asia from the Arctic south to the countries round the Mediterranean, Black Sea and Caspian, and across to the Pacific coasts of China, Formosa and Japan. Their migratory movements, although large and conspicuous, do not take them far beyond their breeding range. Their flight on such journeys is dogged and purposeful but laboured, with spasms of wing-flapping interrupted by longish rests, the air-speed being slow—between 20 and 30 miles an hour. It gives no indication of the tireless energy shown in the quite different song-flight, when the cock rises almost vertically, head to wind, with fanned and depressed tail and deeply winnowing wings, becoming a living organ from which the loud, spirited, shrill warble pours uninterruptedly into the air for three, four, or five minutes or even longer. When the singer reaches his ceiling, perhaps at two or three hundred feet, but sometimes at a thousand or more, his wing-beats become shallower and his flight sometimes circling, although he maintains a strong preference for facing upwind and a remarkable reluctance to wander far from the airspace above his particular patch of ground. Sometimes the descent is made by gradually losing height in the same way, but often it ends in a swift almost vertical dive with feet fully extended to preserve an even keel, the song frequently continuing almost until the bird's feet touch the ground. It is a performance whose overflowing energy and spirit have for centuries made bird-watchers of all its beholders, and whose charm never ceases to give delight.

Although there is no month in which the song cannot be heard it is at its best only between February and July, and is very weak and casual late in the year. Occasionally singing may be heard from flocks; near the southernmost tip of Devon in October I heard one of a flock of about a hundred skylarks start singing, followed by enough others to give for some minutes the illusion of a flock of larks singing overhead.

When flying about the fields the action is quite different from either the song-flight or the migratory flight, and the broad triangular sail-shaped wings are fluttered as if their power were an embarrassment in the short, low, deliberate flights, which often seem to serve for patrolling and looking round, such as other birds might perform from a tree or other point.

On the ground skylarks walk rather stiffly, straining upright or crouching when anxious. They are fond of dusting and I have seen one in a ploughed field on the Marlborough Downs throw up such a thick cloud as to hide itself completely from view; it was later imitated by some other small bird, apparently a linnet, and several house-martins settled momentarily near by, apparently fascinated by the example.

Skylarks eat a wide variety of food readily accessible in the fields, especially seeds and insects. They nest always on the ground, and usually in the shelter of crops or herbage, lining a small hollow with grasses, bents and sometimes hair. Usually the nest (see Pl. 10, p. 49) is in the open, but in Germany I have found one in the middle of an osier bed at the foot of a very small osier. The clutch is small, usually only three or four, and the rounded greyish eggs are thickly speckled and sometimes broadly zoned with dark brown. Laying begins in late April or May, and the young are hatched in about eleven days and leave the nest only nine or ten days later. There are two or three broods. The incubating hen will sit very close; once while making a bird census of a bog in Connemara I was surprised to see a tailless skylark rise at my feet and flutter off, and looking down found that she had left behind ten tail-feathers on which I had actually been standing at the moment that she rose, besides a nest and three eggs which had luckily escaped damage.

The plumage of both sexes is of varying shades of brown with plenty of dark-brown and buff streaks, the underparts and trailing edge of the wing being mainly buff and the longish tail having broad white edges. The crown is distinctly crested. The length is about 7 inches, yet the wings are almost as long as a starling's, and the tail longer.

YELLOW WAGTAIL

Wagtails are brighter, slenderer descendants of the pipit stock, in which the normal pipit browns and buffs are replaced by greens, blues, yellows or greys or black-and-white. The pipit streaks and spots give place to cleaner patterns, the tail becomes longer, the voice shriller and more emphatic, but the song fades away to a feeble and unimportant warble, and the distribution becomes tied to water or to places where special conditions can be found.

The yellow wagtail (see Pl. 11, p. 64) constantly reminds us by its habits and appearance of this affinity between the wagtails and the pipits. Unlike other wagtails, and like the pipits, it is a field-dweller, but it often shows a wagtail's fondness for being near water. Like other wagtails it runs to bright colours, but in the female spots are liable to appear on the breast, and the back may be almost as brown as a pipit's, while the tail is not much longer than a tree-pipit's, and the call-note not much more powerful or more musical. Like the tree-pipit also, and unlike all other members of the family, the yellow wagtail is a regular summer migrant to Great Britain. But the species *Motacilla flava* L. to which our yellow wagtail belongs is distinguished among all other wagtails by its extreme variability. When a party are seen together in spring no two look quite alike, and although blue-headed, grey-headed, ashy-headed, black-headed and other distinct geographical races are recognised it appears that individuals indistinguishable from these forms are liable to crop up as mutants among other races, thus creating a confusion which remains to be fully sorted out.

Although sometimes common where it occurs the yellow wagtail is a bird of very localised distribution, fond of rich, open, lowland pastures or meadows, and certain sorts of cultivated fields. It appears to be basically a marsh bird, spreading over farmland where similar conditions of food and shelter have been created, and also colonising banks of reservoirs, lakes or rivers flowing through open country. In Norfolk I have found it on dry heaths breeding among bracken, although at no great distance from the nearest marsh or stream. In southern Britain it seems to avoid high altitudes, 500 ft. in Wales being the most elevated breeding area of which I have personal knowledge. But R. Chislett has pointed out that this does not apply in Yorkshire, where breeding on high ground is regular. Even in the most favoured areas yellow wagtails will gather at certain places and avoid others. Thus walking down the Isis from the Rose Revived, where the Windrush joins it, I have found yellow wagtails frequent at the start, then none for several miles down to Bablockhythe, then plenty, gradually getting sparser towards Swinford Bridge near Eynsham where they ceased, but reappeared plentifully on the King's Weir and Godstow reaches. Although anxious over human disturbance they often settle close to towns, and during the war they immediately began breeding on the Regent's Park cricket pitches, then occupied by the Army. In June 1925 I found a pair carrying food to nestlings on the bed of the large Queen Mary Reservoir at Littleton in Middlesex, which was shortly afterwards covered with water, and had previously been woodland. Yellow wagtails are fond of perching on fences, posts, low trees or other suitable look-out stations. On passage and after the breeding season they are more widespread, being found on hills, clifftops and other open turfy places.

Although not colonial nesters yellow wagtails are often close neighbours in the breeding season and appear to be by no means strictly territorial. At other times they are the most sociable of our wagtails, although normally met in parties rather than in flocks. They are noticeably fond of accompanying grazing cattle and horses, being attracted, no doubt, by the ease with which flying insects can be caught near them.

The yellow wagtail, like the pied, is a clearly marked British race, spilling over in small numbers as a breeding form to the opposite coasts of the Channel and the North Sea, and possibly to Portugal, but replaced from Norway, Belgium, Central France and North Italy eastwards by the blue-headed wagtail, distinguished in breeding plumage by the white chin and eyestripe and bluish-grey crown and face of the male bird. Other races occupy zones farther north (grey-headed) east (Sykes's), and south (ashy-headed and black-headed), so that our bird represents a relatively small, north-westerly outlier of the species. It is accordingly not surprising that the yellow wagtail should be much commoner in the east and south of Britain than in the north and west, being almost absent as a breeding species from Devon and Cornwall, West Wales, most of West Scotland and from Ireland except possibly in one or two places. Unlike the grey and pied wagtails it withdraws entirely from Britain for the winter, which it prefers to spend in tropical West Africa. Apparently there is no record of even isolated cases of wintering, which is surprising seeing that it is accustomed to arrive in early spring before the country has warmed up, and to remain until the autumn frosts have set in.

The yellow wagtail is a graceful and dainty walker, but nervous and tense in manner. Like other wagtails it is given to making long flights, its action appearing rather less undulating than the pied's, and it more often flies round and round over one area, checking or hovering frequently, and showing little turn of speed. This flight, and the bright

C. W. Teager

Fieldfare

Ralph Chislett

Redwing on breeding-ground in Swedish Lapland

colouring and loud call-note make it one of the most conspicuous species in spring wherever it occurs, and perhaps explain why its song is so feebly developed and so little used, being no more than an occasional snatch or two of low and not unmusical warbling, which may be delivered in ordinary flight and sandwiched between the persistent, characteristic and compelling shrill call-notes, unsatisfyingly rendered "tsweep . . . . . tsweep". These call-notes, once learnt, proclaim the presence of the bird often before it comes in view, their carrying power and continuity of utterance being remarkable. Few bird-sounds except the hunger cry of the young cuckoo have an equal capacity for demanding attention.

The food so far as known consists of insects and small molluscs, taken largely among short herbage. Yellow wagtails are stated to roost on the ground or sometimes in reed-beds, but roosting in rhododendrons has been recorded. The hair-lined nest, built of bents and roots in a hollow of the ground, is usually well screened by thick grass or plants and is not easily found without watching the bird from a sufficient distance to overcome its marked suspicion of human observers. There is however no difficulty in knowing where a yellow wagtail breeding territory begins, since the birds are incapable of skulking or remaining silent, and soon rise to accompany any intruder. They often skirmish among themselves. From four to eight eggs are laid, most often half-a-dozen, smaller and more oval than a pied wagtail's and almost covered with fine flecks, not unlike a paler meadow-pipit's. They are laid usually between mid-May and early June, and take a fortnight or so to hatch. I have seen nestlings being fed recently in the first week of June within four miles of Hyde Park Corner. Both parents are very active in feeding them, and they fledge within a fortnight, a second brood being usual in southern England.

It is difficult to describe the plumage simply, as there is so much variety, but the underparts are always more or less yellow, very pale and broken by a dark-brown band in the young, brighter on the hen and exceedingly brilliant on most breeding cocks, while the nape and mantle are brown more or less strongly tinged with green or yellow or both and the wings and tail are blackish-brown with white or pale buff outer tail-feathers and rather inconspicuous buffish wing-bars. The head of a breeding male varies between bright yellow, greenish or brownish, and some yellow wagtails show slightly bluish-grey on the head, although never of the bright clean tint characteristic of a blue-headed in breeding dress. Identification of the races of yellow wagtail when not in breeding plumage is more readily and safely attempted in the museum than in the field, and puzzling individuals occur, whose origin and subspecific identification yields material for much argument among ornithologists, without seeming to confuse or embarrass the wagtails themselves, who mingle and mate regardless of the colour of their heads. The length of the yellow wagtail is about 6½ inches, but it is in all respects almost as large as the pied wagtail except for its tail, which is nearly a fifth shorter.

FIELDFARE
(Pl. xi, p. 60)

Fieldfares are large, tough, harsh-voiced, skewbald thrushes different in character and habits from all other British birds. Their lives, like their plumage, are a blend of seemingly discordant elements. They are alert and wary, but sluggish in reacting and not difficult to stalk. They like to perch on top of the highest trees and fly at a considerable height, but they feed and roost mostly on the ground. They are far northern birds, and unlike our two other common wintering passerines, the redwing and the brambling, they have never been reliably recorded breeding in the British Isles; yet hundreds of miles south of us, in Switzerland, they breed locally and are extending their range. They are probably our most characteristic hard-weather birds, yet they may arrive as early as August or September and not leave until May, or even June or July. One May Day near Selborne I saw a flock of sixty or seventy fieldfares fly over a hedge in which young mistle-thrushes were already moving about fully fledged. Swifts and turtle-doves had already arrived.

Fieldfares belong to the fringes between open and wooded country. They need big fields or grassy wastes, dry or damp, high or low-lying, but with easy access to tall trees. The berry crop, however, will sometimes tempt them to settle (while it lasts) in tall dense thickets of thorn, yew, holly and other mixed trees and shrubs, and they will sometimes settle well inside woods, although not normally in the undergrowth or lower branches. Large blocks of woodland, orchard, gardens away from the country, and sedge, reed, heath or moor are generally avoided. They quite often appear in towns, even in the central London parks on passage or in hard weather, but usually move on elsewhere almost at once. Although they like open ground, very large exposed areas far from cover do not much appeal to fieldfares. In Sweden in late summer I found them common in spruce and in birch up to about 2,500 ft. on the fells. They breed, unlike our other thrushes, sociably, usually in woods, but often in town gardens.

From the arctic tree-limit across northern Europe and most of Asia fieldfares breed southwards through the forest belt, with southerly outliers in the Carpathian and Alpine mountain forests, and with occasional nests even in Holland and eastern France. Although there is a general southward movement in winter it takes them only a few hundred miles, and few penetrate as far as the Mediterranean. Unlike redwings, fieldfares do not breed in Iceland. They visit all parts of the British Isles, but in very varying numbers in different seasons and weather conditions, being usually a good deal less numerous and more patchily distributed in the south than redwings, although at times they become the most numerous species of thrush.

Isolated fieldfares are unusual, single birds or small parties generally proving to be temporarily detached outposts of a large loose nomadic gathering, which may typically have a strength of one or two hundred fieldfares and at least as many hangers-on in the shape of redwings, starlings and other birds. At times very much larger flocks are seen. When feeding, fieldfares sprinkle themselves rather thinly over the ground, usually all facing the same way, even if there is no breeze. They move in brief quick rushes. If disturbed they often get up in ones and twos or small parties and trickle over to the next field or to some watch-tower in the crown or fringes of the most commanding trees. There they sit in their stiff almost wire-drawn posture, appearing never nimble and never relaxed. Sometimes on perching the tail is cocked, and I have seen one perch with its back to a stiff breeze.

Fieldfares fly rather like bee-eaters, with strong but easy strokes interrupted by pauses just enough to break the rhythm without introducing the bounding up-and-down pitch of the mistle-thrush, although when alarmed or coming in to perch the flight does become quite undulating. They are fairly fast fliers, and will rise to 300 feet or more for any but the shortest movements.

Slugs, worms, insects, and sometimes seeds, grain or roots, are picked up in the fields, and a very wide variety of berries on the higher shrubs and trees. Quartering themselves in large numbers within a small area fieldfares need some sort of abundant food supply which is there for the picking, and this requirement no doubt accounts for their patchy occurrence and nomadic movements.

It seems doubtful whether the fieldfare can be said to have a song, its performance being described by the *Handbook* as "very feeble, with some notes suggesting a poor Blackbird, mixed with chuckling, whistling and harsh squeaky notes and variations of the harsh call". It is uttered chiefly in flight. Living and breeding sociably fieldfares have had no cause to develop the full song for which they undoubtedly have the physical capacity. Even this sub-song is heard in England only rarely and in a yet feebler version. The normal sound with which fieldfares announce their presence from afar is a loud chuckling, grating, chacking call-note, uttered several times without a pause. Besides this chatter, unmistakable once learnt, there is a rather wistful soft drawn-out note not unlike a redwing's.

Although slightly smaller than mistle-thrushes fieldfares are similar in build and proportions, but their peculiar bearing and variegated colouring give a very distinct impression. The soft french-grey of the head and nape and the rump contrasts with the ruddy chestnut back and the black wings and tail, while underneath the thick bold spotting of the upper breast is set off by the rich buff ground-colour and again contrasted with the white lower breast and belly, and in flight the underwing. The effect is as if a conventional brown-backed spotted breasted thrush had undergone some strange mutation half-way towards a brilliant plumage pattern, and there remained, neither one thing nor the other.

REDWING

(Pl. XII, p. 61)

Just half-way round the world, from near Kamchatka to Iceland, stretches the breeding range of the redwing, which follows the zone of birch and alder scrub and forest across Arctic Asia and Europe, with a depth in most parts of less than a thousand miles. Few nest south of the Baltic, and the breeding range misses the British Isles by about 200 miles, coming as near as Norway. Iceland and Faeroe are inhabited by a separate larger and darker race. In winter both sub-species invade the British Isles where many remain until spring; the migration is nowhere very distant, stopping short before the Mediterranean. On at least three occasions pairs have built nests in

Eric Hosking

Yellow wagtail and young in a Suffolk marsh

PLATE 12

H. M. Stone

Little owl

Scotland, and one pair, thought to belong to the Iceland race, actually reared young on Fair Isle in 1935.

Redwings scarcely give the impression of being suited to their hard northern life. Meek and subdued in manner, unobtrusive and delicate, they are amongst the earliest and heaviest sufferers from any spell of hard weather, and seem much less fitted to stand an English winter than our ring-ouzels which never face it, or even than our throstles, many of which emigrate until spring. They often mingle with fieldfares. They are, however, more attached than fieldfares to tall bushes and to trees of no great height, and they have a stronger dislike of getting too far from cover, although nothing like so determined to keep within a short dash of thick shelter as our own blackbirds and throstles. They also show a more definite preference for good grassland, and for smaller and less rough fields, and are fond of parks and playing fields, although they will not face much disturbance. These preferences are no doubt related to their diet, which contains little vegetable matter except for a gorge of berries when opportunity offers, and is mainly composed of insects, worms, slugs, snails and other small creatures. Floodland often seems to attract them, perhaps because it offers more accessible food resources. For roosting they like to retire to some of the thickest and warmest cover in the neighbourhood, usually in a wood, shrubbery, or plantation, where they assemble in quite large numbers, and sometimes mingle with fieldfares, starlings and other gregarious sleepers. One roost which I found on the Chilterns was in the heart of quite a large tall beech-wood. After feeding all day on the Thames levels the redwings would flight to it in bands of a dozen to seventy or more, the movement, which I could see regularly every afternoon, being spread over about half-an-hour. At the beginning of February I estimated that there were at the very least a thousand birds in it, and in March there were considerably more. On arriving at the roost a whole swarm of them would whirl round starling-like in a wide circle; on dropping into the trees they became silent; they were shy and unapproachable.

Flocks on the ground feed in very open order and in silence, often moving little unless disturbed, when each individual shifts as it happens to notice the danger, and usually sits up in a near-by tree, often staying to balance on the topmost twigs, or slipping through to the far side from which it retreats without giving much of a view. When on the move redwings travel quite high but are recognised without difficulty by their characteristic action, which is more lark-like than thrush-like, and by the prolonged plaintive far-carrying "seep" flight-note, so often heard from flocks migrating over us in the dark as we lie awake some October or November night. The flight appears slow, and gives an impression of struggle when there is an adverse wind, the birds coming down to treetop level to snatch what relief they can from it. Flying flocks look straggling and disorderly, especially when some obstacle is met, which they only seem to see and alter course round at the last moment.

The alarm cry is a rather harsh and grating "krek-kek" less loud than the fieldfare's and less used at most seasons. There has long been argument about the redwing's song, and whether it is uttered in this country. As far as is known there are three distinct performances which might be classed as song. The fullest and most advanced version is described as being much like the throstle's but rather lower-pitched with many phrases each repeated several times. This apparently is confined as a rule to the period before incubation. The second version is a single phrase of a few clear fluty notes, also heard on the breeding ground. The third version, which is probably best treated as sub-song is the only one which is definitely heard in this country, and is in fact common in February and March. It is a low continuous babbling warble full of harsh and even screeching or croaking sounds, mixed up with much pleasant twittering and musical conversation, more like a starling's song than any other. It is often delivered in chorus by a number of birds, and the effect sometimes when a hundred or more take part is very pleasing. It seems to become more frequent and more intense as the birds' departure approaches. At that time they become increasingly restless, sitting in parties in the thick of the trees and babbling to one another. They leave earlier than fieldfares and are normally all gone by about mid-April, although an occasional straggler lingers till May or even June. Early arrivals have been recorded from July to September but the main immigration is in October and November.

Unlike fieldfares redwings nest singly, usually in trees or bushes but often on the ground, and they manage to rear two broods.

Seen side by side on the ground the throstle is a considerably bigger bird than the redwing, standing conspicuously taller and with a more golden brown back. More striking distinctions are the redwing's chestnut flanks and axillaries with a strong vinous tinge—the German

"Weindrossel" is a less misleading name, for the bird shows no true red and the colour is mainly on the flanks and tucked under the extreme base of the wing. On the ground the long broad buffish-white streak above the eye is conspicuous at a good distance and is sufficient to pick out redwings of both sexes and all ages. The wing is proportionately rather longer than the throstle's.

LITTLE OWL

Charles Waterton, among his many eccentricities, seems to have been the first to get the idea that it would be a good thing to introduce the little owl (see Pl. 12, p. 65) into England. Finding it "much prized by the gardeners in Italy for its uncommon ability in destroying insects, snails, slugs, reptiles and mice" he jumped to the conclusion that it would be a benefit to have it here. He therefore brought a dozen from Italy in 1841 but he gave them a warm bath at Aix-la-Chapelle and five of them "died of cold the same night". Five survivors were released near Wakefield in south Yorkshire on 10 May 1842. This attempt failed, and being wise after the event we can see that the site was at the very northern limit of suitable conditions for little owls in England, even if there had been no other adverse factor. But ideas once started can be as hard to eradicate as little owls themselves, and about thirty years later E. G. B. Meade-Waldo took up the game more thoroughly and on more promising ground at Edenbridge in Kent. By 1879 the first pair bred, but it was only after further introductions that he was able to record regular breeding beginning in 1896. Meanwhile Lord Lilford had imported and released large numbers near Oundle, in Northamptonshire and found his first pair breeding in 1889.

Other introductions were mostly abortive, and it was from Oundle and Edenbridge that the big spread began in the 'nineties. By about 1915 little owls had colonised much of the area east of a line from Southampton Water to Cardiff and thence up to the Humber. In the next decade the expansion seems to have reached its height, with little owls pushing south-west to Land's End, west to Pembrokeshire, north-west to Caernarvon and Lancashire, and north and north-east to mid-Yorkshire. Not only were new areas being colonised, but densities appeared to be increasing in the regions settled earlier. After 1925 a few pairs bred in Durham and Northumberland, and odd birds turned up in Cumberland and Westmorland, but these last counties, together with Anglesey, and the whole of Scotland and Ireland remained uncolonised. While increase continued during the 'thirties in some districts on the fringe of the range there were many reports of decrease from the older settled areas, including parts of Kent, Hampshire and the Isle of Wight, Hertfordshire, and Cambridgeshire. Some spread into Durham and Northumberland has since taken place.

As the wave of little owls rolled across England a wave of hysteria sprang up in its wake, and eventually overtook it. Aristotle nearly 2,300 years ago correctly described little owls as hunting "in twilight and at early dawn. They hunt mice and lizards and beetles and such other small animals". Waterton a century ago had taken the same view, and from 1918 onwards Dr. Walter E. Collinge examined more than three hundred little owl gizzards and nearly as many pellets, again finding that insects, voles, and mice constitute the chief items of the little owl's food throughout the year. Yet throughout the 'twenties and 'thirties a number of gentlemen whose powers of observation and logic would have done them more credit 2,300 years before Aristotle than 2,300 years after him carried on a virulent campaign of emotional abuse against the little owl, which they pictured as emptying our coverts of game and our hedges of songsters. In 1935 the gulf between a section of opinion and the available scientific material had become so wide that the British Trust for Ornithology was asked to promote a comprehensive new inquiry, which was carried out with the greatest thoroughness, care and accuracy by the late Miss Alice Hibbert-Ware of Girton under the scientific supervision of a very eminent Special Committee consisting of Dr. Collinge, Mr. (afterwards Sir John) Fryer, later Secretary of the Agricultural Research Council, the Rev. F. C. R. Jourdain, one of the most exacting and best-informed judges of ornithological work who ever lived, and Mr. N. B. Kinnear, later Director of the British Museum (Natural History). A general investigation was carried on for a year from February 1936, and a special investigation was added during March-July 1937 in order to secure fuller material from places where game and poultry were reared. Critics of the little owl were throughout invited to take part and put in evidence of its reported depredations. Food material was collected from every English county where little owls occur regularly, except four. In addition much intensive and extensive field observation was done, and experiments were carried out at the Zoo and elsewhere.

T. M. Fowler

Barn-owl returning with vole

T. M. Fowler

Young barn-owls

This inquiry showed that daytime hunting was much less usual than night hunting, and that insects and rodents taken on the ground were the prevalent food at all seasons. Carrion was little used and there was no evidence to support stories that prey was killed in excess of immediate needs, either for storage or for use as a breeding ground for carrion beetles to be eaten later. During the nesting season rats and rabbits were taken, and voles and mice, sometimes in abundance. Starlings, house-sparrows, blackbirds and throstles were also taken quite commonly in the breeding season, but other birds, including game chicks only rarely, except by occasional aberrant individuals or in very unusual circumstances, as when a pair on a remote island took to feeding on storm-petrels. There was no evidence of birds being taken from bushes or trees, or of nests being raided (except occasionally nests in holes), or of egg-sucking. Large numbers of cockchafers, crane-flies, earwigs, dor beetles, carabid beetles, and weevils were eaten.

The Special Committee fully endorsed this report, finding "proof of only negligible destruction of game, poultry, or wild birds of all ages" but commenting that "during the period of rapid multiplication of the species, which seems to have come to an end some years ago, there may well have been local tendencies to depart from the normal diet owing to greater competition for food or a relative lack of the kinds usually preferred".

Notwithstanding the repeated broadcast and Press invitations for those having evidence against the little owl to bring it forward this patient and comprehensive investigation found nothing to substantiate the wild and sweeping condemnations which had been passed on the little owl. But prejudice dies hard, and although the verdict of the Inquiry is increasingly widely accepted there are and will no doubt long continue to be plenty of ordinarily sober observers ready to repeat unsubstantiated statements and to discard science in favour of emotion when the little owl is mentioned.

Reports from various parts of the country showed that in 1936–37 areas favoured by little owls contained from one to four pairs within a half-mile radius, although in some cases the density had been a good deal higher previously. They prefer rather open park-land, farmland or sometimes orchards, with good cover such as hollow trees, quarries, and sometimes buildings.

Little owls, unlike tawny owls, do not nest in inner London, but they are sometimes heard passing through. They are especially fond of perching on low posts, fallen trees, wire fences or telegraphs or power lines, and pollard willows or other lines of trees on the edge of open ground. They seem rather to avoid woods. One which I once watched sitting on a clod in a rabbit warren on the flank of Birdlip Hill in Gloucestershire suddenly vanished down a rabbit burrow and refused to be dislodged. This was in March, and the bird may have already begun nesting there. Normally breeding does not begin until May or nearly, in almost any sort of hole, although rarely in an inhabited building. Early this century four or five eggs were laid regularly and quite often more, but recently three or four, or even only two eggs have become more usual. They are dull white and hatch after 28-29 days, the young fledging in about 26 more; a second brood is certainly not usual.

Although often sitting still, little owls are alert and quick in their movements, and can run fast on the ground. The normal flight is from about ten to about forty feet up, with a series of strong wing-strokes alternating with longish pauses to produce a conspicuously up-and-down motion, but sometimes the action is simply flapping and gliding without the least undulation. When alarmed they bob comically up and down. They sit very upright.

Little owls are on the whole noisy birds, both by day and by night, calling to each other with a variety of notes which have little of the mysterious and unworldly quality of many owl voices, and tend to become either brief yelping or petulant challenges and responses or harsh rather more prolonged choruses.

In appearance the little owl is conspicuously flat-headed, with a permanent frown as if of surprise and distaste implanted on its shallow facial disks, and with big yellow eyes, a plump body and a broad, short, rounded tail. The length averages under 9 inches. The plumage is less soft than in the other common owls and is all in two main tints—a dark greyish-chocolate brown background with numberless whitish or buffish-white spots, bars and mottlings.

BARN-OWL

Almost all round the world barn-owls come out to hunt as the sun passes on to more westerly lands. Different races represent the species in different continents and regions, from Europe through India and Australia to the Pacific Islands and North and South America. But despite their warm downy plumage they avoid the colder regions of the earth, being absent from Siberia and North Russia, much of Scandinavia, and most of Canada, and in the south from the Antarctic, Patagonia and even New Zealand. Nowhere else on the globe do they reach out so far from the Equator as in the Highlands of Scotland.

Our own white-breasted race of barn-owl (see Pl. 13, p. 68) is the typical subspecies, and is mainly a Mediterranean form, stretching from Asia Minor and Iraq through Egypt to Morocco and the Azores and up through Spain, West France, Sicily and Switzerland to Belgium and the British Isles, while the dark-breasted race takes over from south and east France, Germany, Holland, Denmark and South Sweden eastward.

Although so widespread the barn-owl seems to be nowhere very plentiful. In the British Isles, it is found in all parts except the northern Highlands, Orkney, Shetland and the Outer Hebrides. Last century and earlier it was a familiar and in many parts a frequent species, but recently its numbers have sadly declined, and it often now ranks behind the tawny and little owls, and locally even behind the long-eared, which is commoner in Ireland and parts of Scotland. G. B. Blaker, who carried out an enquiry into the barn-owl's distribution and numbers in England and Wales in 1932, reported a curiously patchy distribution, with the least thin densities in Cumberland and Westmorland, Anglesey, parts of Devon and neighbouring areas and parts of Essex. The species hardly lends itself to reliable census work at the present stage of ornithology, but enough evidence was gathered to indicate that except in the extreme north of England and in and around Devon general and substantial decreases had been taking place for some years. Some observers believe that this trend has lately begun to be reversed, but there is as yet no conclusive evidence.

No one would seriously argue that barn-owls are harmful, and most people who are in contact with their ways regard them as beneficial. They have long been legally protected in most counties, and no wild enemies normally attack them. That a bird in this situation can nevertheless fall into a decline makes plain how ineffective our good intentions and efforts as a community at present are to achieve their objects, and how hopelessly we are in the dark about what makes even some of the birds nearest to us flourish or fail. The proscribed and black carrion-crow multiplies; the protected and white barn-owl fades away.

We know, of course, that regardless of the law large numbers of barn-owls have continued to be shot every year, sometimes for trophies or firescreens, sometimes because of complaints of their snoring and weird night disturbances, often simply out of wanton destructiveness. In April 1947 thirteen dead barn-owls were counted on a vermin pole in north Lincolnshire. There is some evidence, also, that some barn-owls may be poisoned through eating rats and mice which have themselves taken poison but not yet died of it. As the barn-owl in Great Britain is so near the northern limit of its range, and is a sedentary species, it is possible that the cumulative effect of the severe weather early in 1929 and in several more recent winters has been a factor in depressing its numbers. Ringing returns over a period of years, however, show a percentage of recoveries of birds aged less than two so heavy as to point strongly to exceptional mortality as a regular feature. The odds against a barn-owl dying of old age seem pretty heavy. There may well also have been some correlation between the decline of the barn-owl before 1939 and the shrinkage of cereal cultivation, operating through the supply of mice. There is some indication that barn-owls are densest either in corn-growing districts or in areas with plenty of rough herbage affording cover for voles, while most areas of improved grassland seem less attractive to them. If food has acted as a limiting factor it may well have led to a lowering of the average brood size, since the fertility of owls is very responsive to food conditions. These suggestions, however, must be regarded rather as indications deserving further inquiry than as findings based on adequate information.

Barn-owls are attracted to the neighbourhood of people partly by the many suitable dark breeding places provided by farm buildings, roofs and chimneys of houses, church towers and ruins, and partly by their depending so largely on small mammals which are often commoner and more easily caught on cultivated land. They also nest freely in hollow trees, sometimes in preference to farm buildings near by, and where owl-boxes are provided they may use these; dovecotes have often been invaded by barn-owls for the same purpose. Occasionally, however, they break right away and make their homes in a remote quarry or crag up to 1,000 or even 1,500 ft. above sea level, with such birds as ravens for neighbours. A few nests are placed in dark spots in trees without the protection of a hole, especially in yews. In France I have seen a pair passing to and from their nest on the cathedral in the main square in the middle of a small city, but in British towns the tawny and the little owl are much more usual.

From these bases barn-owls go out to hunt over the neighbouring open country, sometimes singly, sometimes in pairs. Although normally nocturnal they may not infrequently be seen hunting by day, especially on winter afternoons or summer evenings when food is either harder to get or is needed in bulk by the young, and sometimes in other months. One April afternoon in the Severn valley I saw one fly out of a wood about 2½ hours before sunset and make its way along a hillside till it met another, probably its mate, and both settled side by side and in full view on a clump of brambles. One soon flew away and began hunting, ranging over ground as much as half-a-mile off. It took no notice of me, flying within five yards or so as I stood still, the curious peaked face being conspicuous as it approached. When it pounced, as it often did, it dropped lightly and remained with its wings stretched out loosely on the ground but before taking flight again it seemed always to fold them. The legs seemed to be used as a rudder and their position was often shifted in flight, while the tail supplemented the supporting surface of the long broad wings. They were silent, but seemed thoroughly at home by daylight during the two hours that I watched them.

Another time in the same district and also in broad daylight, I put up from a roadside ditch a barn-owl which was promptly pursued by a crow or rook. It alighted and its pursuer settled alongside, to resume the chase when it flew again, the owl finally shaking off pursuit by passing close to me and taking refuge in a hedgebottom, from which it afterwards extricated itself clumsily and with unaccustomed noise. Normally in hunting the flight is not only noiseless but remarkably slow and quite near the ground. Shrews, long-tailed field-mice, field and bank-voles, house-sparrows and starlings are the commonest prey in most places. Rats and house-mice are only occasionally taken, as are bats, rabbits and other small mammals, chaffinches, swallows and other small birds, moths, frogs and fish. Beetles are eaten fairly freely. After feeding, the owl coughs up such indigestible items as fur feather and bones in a neat compact pellet, in which they remain clearly recognisable for long periods—the pellets of the barn-owl are distinguished from those of other owls by their blackish colour, smoothness and rounded shape. They are cast up at least twice a day at regular perches, from beneath which they can be collected for analysis.

Sometimes the evidence of pellets shows that a roost has been shared with another species, and it is on record that a barn-owl and a little owl have been flushed from the same place and pellets of both found in the same hole. I know one place in Devon where tawny owls regularly roost in an old ilex close to an ancient church tower which is the stronghold of the barn-owls, apparently without friction between the two.

The language is varied and weird, the commonest notes being a prolonged shriek, and a loud snoring. Several good observers have recorded barn-owls hooting, but this is evidently exceptional. They are more silent by day than tawny and little owls.

Any number of white eggs from 3 to 11 are laid at unusually long intervals and in nearly any month of the year, although oftenest in April-May, a minority rearing two broods in the season. They take over a month to hatch and the young (see Pl. 14, p. 69), which are often at different stages of development do not fly for two or three months longer, making a very protracted breeding-season.

The plumage is beautifully soft and satiny, and the very pronounced pure-white heart-shaped face is like no other bird's. The underparts are white too, and the long legs, on which the bird stands fairly upright, are covered with white feathering, while the upperparts are golden-buff with a good deal of fine mottling, which tends to be heavier in the female than in the male. The eyes are unusually small and dark for an owl. The length of 13½ inches with wings 10-11 inches long are very close to the measurements of that most graceful of falcons the female hobby, but the bill and legs are much longer and larger and the tail considerably shorter, while the looseness of the plumage and the rounded head, wings and tail give a totally different effect.

LAPWING

Lapwings, like farmers, are tied to the soil, and can find their living wherever soil covers the underlying rock without itself becoming buried under too tall and dense a mass of vegetation. Being drawn to the same places by this common tie, lapwings have found themselves living more and more on farms, and farmers on lapwing territories. On the whole it has worked out well, the farmers having hitherto shaped the land near enough as lapwings like it, while the lapwings have busied themselves in eating those creatures and plants which the farmers wish to see (or do not mind seeing) eaten. The friendliness thus inspired in the farmers in turn benefits the lapwings, which although themselves good to eat and laying very palatable eggs are protected not only by a special Act of Parliament but by the even more powerful goodwill of the farming community, without which they would have to fear widespread destruction.

This pleasing but precarious partnership has arisen out of traditional methods of farming which are now undergoing drastic and rapid changes. Whether these changes will further benefit the lapwings, or whether they will inflict more or less serious injuries on them is a very real question, and one which will soon be answered. Lapwings breed on the ground (see Pl. 15, 16, pps. 76, 77), largely in cultivated fields, and the more often and quickly these fields are rolled, harrowed or otherwise disturbed in spring the more of their eggs and young must be destroyed. The vast array of tractors and mechanical implements which have recently transformed the British farming community into one of the most highly mechanised bodies of men in the world now enable our fields to be tended with a speed and heavy-handed thoroughness extremely inconvenient for the protection of lapwings. Unless more reliance is placed on late layings to make good the damage it seems likely that those breeding on arable land will have difficulty in maintaining themselves from now on. At the same time the permanent pastures, which just before the war broke out in 1939 ranked equal with arable as the most favoured breeding habitat for lapwings, have since been marked down for ploughing, cropping and reseeding, and vast acreages have already been given this treatment. Now along comes the policy of frequent cutting of short grass for drying or silage in place of letting it grow long for hay, and at the same time the rough and soggy lowland pastures so much liked by lapwings are systematically reclaimed and drained to yield more and better grass. It appears that the interests of lapwings are being threatened all along the line, and it is difficult to avoid the suspicion that such factors have already played some part in the very marked decline in the numbers of lapwings during the past ten years, although the run of severe winters must also have caused heavy casualties. The large afforestation programme will also render many lapwing territories untenable.

Lapwings, however, have proved themselves over many generations remarkably adaptable to new agricultural processes, and although much anxiety must be felt for them it is too early to regard their long association with British farming as doomed to cease. Certain trends, such as the big increase in ploughing and the tendency to throw fields together by rooting out hedges are actually favourable for lapwings—it is unusual to find them in small hedged-in fields. While the swallowing up of open land for building has driven lapwings farther out from many towns some changes such as the spread of airfields have provided improved conditions for lapwings, although the development of high-speed aircraft makes airmen increasingly concerned to keep flocks of largish birds as well clear of take-off and landing approaches as possible. Like the persecuted Covenanters, who hated the lapwing for giving away their hiding-places, airmen have good reason to follow the Old Testament in classing the lapwing among fowls which are an abomination.

At all times of the year lapwings are fond of the margins of accessible fresh water and of marshes, sewage farms and mudflats, where flocks often gather. Tidal beaches or shores are frequented much less often, and chiefly in winter; one Christmas Day I flushed a flock from a rocky beach near Minehead, but as a rule only muddy or sandy beaches are favoured. Bare stony mountains, and steep slopes generally are avoided, and although sometimes breeding up to nearly 3,000 ft. lapwings seem reluctant to go above about 2,000 ft. or in some districts lower. Some lowland-bred lapwings flock on high ground in the late summer, but in winter the higher altitudes are deserted; at all times moorland habitats are among the least favoured. Golf courses, playing fields and racecourses often attract lapwings, but they avoid enclosed and wooded areas, including parkland with many trees, and they give a wide berth not only to towns but usually to even the most isolated dwellings. Farmland, except when under tall standing crops, is by far the most important lapwing habitat over the country as a whole. Lapwings are unconcerned about passing road, railway, or even air traffic, but are sensitive to people or dogs crossing their territories in the breeding season. Richmond Park, in which lapwings had ceased to be resident from 1835, was recolonised in 1943 through the combined attraction of wartime cultivation and the exclusion of the public. It will be interesting to see how long the birds will maintain themselves here in face of increasing disturbance and the abandonment of cultivation. To some extent the attraction of cultivated land for breeding purposes may be due to the protection which it affords from disturbance, or rather did afford before the recent outburst of mechanisation.

Robert Atkinson

Lapwing incubating

Robert Atkinson

Lapwing's nest in dredge corn; Hampshire

I have often suspected that areas much frequented by rooks are regarded by lapwings as unsuitable breeding territories, and as breeding lapwings feel it necessary to fly up and persuade any kind of visiting crow to move away from their territory as fast as possible it would evidently cause incessant trouble if lapwings tried to breed along a main flyline from a rookery or on a favourite rook feeding-ground. Some cases have been recorded where the two appear to settle down side by side, but further study of this point is desirable. If incompatibility is confirmed it would follow that the increase in rook numbers which has certainly taken place since about 1930 is probably a further appreciable factor in the decrease of lapwings. The general increase and spread during the war of carrion-crows, magpies, jays and probably jackdaw would add to the pressure, which may in any case be seriously aggravated when lapwing numbers fall below a certain level, and there are fewer neighbouring pairs to support one another's defensive efforts.

Breeding lapwings are vigorously territorial, the size of territories indicated by Eliot Howard's plans in *Territory in Bird Life* ranging between roughly half an acre and one-and-a-half acres. At times in very favoured areas as many as a couple of dozen or more pairs will breed on adjoining territories, but even where lapwings flourish most the density over large areas is fairly low. On typical North Cheshire farmland, A. W. Boyd found in 1930 an average of 18 breeding pairs to 1,000 acres, and on 224 square miles covered for the Oxford Rook Census in 1928 I estimated only between 2 and 3 pairs per 1,000 acres. M. D. Lister on a 225-acre Surrey farm found normally 6 to 8 breeding pairs, in a specially favoured area, before the war. Since then, and especially after 1947, lapwings have become much scarcer, and it has been possible to make long journeys across England without seeing any. In one very favourable part of Cornwall in September 1948 I saw more than eighty species including several rare waders without meeting a single lapwing. Later there was evidently some recovery, but it remains to be seen whether the full strength of the 'thirties will be regained.

Few birds are so widely distributed over the whole of the British Isles as the lapwing, which is as much at home on the remoter islands and in the wild hill country as in the Home Counties. Eastwards its breeding range stretches right across Europe from the Arctic to the Mediterranean and Black Sea, continuing across the Urals over much of northern Asia. Without being long-distance migrants lapwings are great travellers, wintering south as far as Madeira and the Canaries, northern Africa, India and Indo China, and east to Japan and Formosa. The most remarkable known journey was the mass transatlantic flight made by flocks of lapwings to Newfoundland in December 1927 with the aid of an easterly gale. As lapwings are not established in north America it would have been evident that these had achieved at least the 2,200 mile crossing from Ireland even without the fortunate confirmation provided by one bird which carried across a ring, number X 5046, placed on it as a nestling at Ullswater in Cumberland in May of the previous year.

A careful reconstruction of the circumstances by H. F. Witherby points to several hundreds at least having been impelled after three days of very frosty weather to leave Cumberland soon after dusk on 19th December, on a course slightly south of west which would normally take them to Ireland. It happened that a fifty-five mile-an-hour gale carried them quickly clear of Ireland, and combined with their own air-speed of about 45 miles an hour took them across direct to Newfoundland in about 24 hours elapsed time, so that they straggled in on the afternoon and evening of the 20th at various points along the coast. They were thin, tame and tired, and in no condition to stand a Canadian winter. Other lapwings have reached America at various points from Greenland and Baffin Island through the Canadian and United States mainland to Barbados, and also on the Pacific side of Alaska.

The British Isles fill several roles in the lapwing's world. For many they are simply a summer home, the winter being spent in France, Spain or Portugal, while many from Scotland and northern England move over to Ireland for the winter. Lapwings from the southern half of England are more inclined to stay at home all the year, and less inclined, if they make any journey, to leave the country. Many birds from across the North Sea and from the Southern Baltic countries cross our east coast in autumn, some staying here while others cross the sea again to Ireland or south-western Europe. Lapwings seen in southern England late in the year may therefore be either local birds, birds from other parts of Britain or birds from the Continent, while at the same time many of the birds bred in the district may be in Spain and Portugal.

The lapwing's year in Britain has three roughly equal phases. In the first, from February/March to June/July, our native birds are at home and almost all on breeding territory. Overlapping this phase from June on is the second phase of summer flocking by the same birds and their young who gather into bands of about 50 to 250 birds, shifting and fluctuating in strength, but usually not going far from the breeding areas, unless these are very inhospitable. The third phase, also overlapping, begins in August/September and is in full swing from October to February. Now all is movement and confusion, with large numbers simultaneously leaving, passing through and arriving in Great Britain at various points and in various directions, while others assemble in flocks numbering several hundreds or thousands. Early in the year a movement of pairs away from the flocks and back to breeding territories completes the cycle.

Lapwings are among the most attractive of birds. Their plumage is rich in colour and glancing with subtle sheens, but at the same time it has the bold simple contrast of an oystercatcher's or a magpie's. On the ground their shape is slender, their attitude statuesque, and their movement graceful. On the wing they become transformed into broad-winged, massive birds, rhythmic in action and capable of reckless swerves and plunges. Handsome as they are when seen singly lapwings have a natural talent for grouping themselves both at rest and in flight so that the form of the flock enhances what each bird brings to it. Their voices and cries have a haunting romantic quality. Their movements are alert and decided, but hardly ever hurried or ungraceful; they are vigilant but not obsessed by nervous anxiety; and aggressive only in self-defence and against birds bigger than themselves. Although profoundly sociable they are full of individuality. Few things in bird-watching are more pleasant and rewarding than watching lapwings.

Lapwings do not allow night and day to tyrannise over their lives, and may be found either briskly feeding or fast asleep at any hour of the twenty-four. When in flocks they spread out widely to forage and draw close in together to rest or preen. Unlike many sociable birds they do not tend to work over the ground in a moving band, but having settled on a field quarter it to and fro facing in all directions, unless the wind is uncomfortably strong. They walk or run a few yards, then stop, and pick at the ground, sometimes with a smart stab at something on the surface, sometimes with a more complex movement as in dragging out an earthworm, occasionally also snapping at flying insects. Insects and other invertebrates form the bulk of the diet but some seeds and vegetable matter are also eaten.

In the fields rooks and jackdaws, starlings, mistle-thrushes, black-headed and mew gulls, curlews, oystercatchers and golden plover, and even at times moorhens, pheasants, mallards and other species may be seen feeding mingled with or alongside the sociable lapwings. Except with the starlings flocks of lapwings seem to dislike letting this mingling go too far, and brief brushes with other species are frequent.

In some places gulls have taken to making lapwings give up food to them. Nearly all these field birds will often perch on some tree or bush, or at least some post or wire or haycock overlooking the field but, although a lapwing has been recorded doing this, you may watch for a lifetime without seeing another do it. It is most unusual for lapwings to show any sign of recognising the existence of anything sticking up from the ground, unless giving a wide berth to such objects is a form of recognition.

There is some evidence that lapwings avoid fields where poultry are kept, but they certainly do not avoid other livestock such as cattle, sheep and horses, and on the Berkshire Downs I have seen some members of a large flock running about feeding in a crowded sheep-pen almost under the feet of the folded sheep, which were busy eating down growing clover. Such confined spaces are rarely entered, but in Galloway I have seen some visit, with some hesitation, a 10-yards wide pool in a stone quarry with a 10-foot high rock-face on one side and a 4-foot grass bank on the other. This was used for bathing or drinking or both.

Compared with nearly all other birds, and particularly with most of its relatives among the waders the lapwing has very broad rounded wings, suggestive of soaring habits. Yet it never soars, and does not do much gliding, using for long-distance flight a rather slow, deep, stylised wing-beat, faintly reminiscent of a small heron. The characteristic up-and-down action and the way the black-and-white underwing catches the eye enable the movements of individuals in a wheeling flock to be followed more distinctly than in other cases, and give an exaggerated impression of lack of co-ordination and raggedness in their evolutions on the wing.

In display-flight the male goes out of his way to look clumsy in the manner of an acrobat who is always about to miss the next one but never actually does. He towers and plunges steeply, with drunken

PLATE XIV.

Eric Hosking

(*a*) Red-legged partridge

John Markham

(*b*) Quail (in captivity)

PLATE XIII.

Robert Atkinson

turns and twists and tilts, making with his wings a loud humming throb which has been aptly termed "bull-roaring". At sunset one December afternoon in the Thames valley I saw a mass display flight by a very large flock, part of which quartered up and down within thirty feet of the ground plunging at intervals with a soothing roar of wings like a wave breaking on sand on a calm night; then the whole flock drew apart and the two divisions, opening ranks, flew right through one another without warning and at some speed. I distinctly heard the impact of five or six collisions in this bewildering aerial countermarch.

Lapwings often fly at around 30 miles an hour, but on migration or when pressed they are capable of between 40 and 50; they will rise to great altitudes, and have not infrequently been recorded a mile or more high.

Much display takes place on the ground, where various scrapes are made well before nesting begins; the male makes himself conspicuous by up-ending with his brilliant pinkish-buff under-tailcoverts rearing up like the breast of some great robin as he holds his tail almost straight upright, dips the breast in dancing rhythm down on the ground, and holds the glossy-green, sharply-pointed folded wings erect well apart from the body, all the while working hard with his feet. The most elaborate ceremony I have ever seen was as late as 20 May 1939 near Pulborough in a wet watermeadow, where a presumed male, after some bull-roaring flight with the characteristic *willock-a-weet* or *peerrweet-weet-weet* cry alighted in full view within 30 yards of me and settled down, in the presence of the presumed hen to lower his body and head and raise his tail, with wings loose, constantly uttering a gruff low note not unlike a starling's alarm but less harsh; in the next phase the breast was sunk right on to the ground, the bill apparently touching it, the wings raised and opened showing black and white feathers, and the tail depressed at right angles to the wings and violently agitated in a kind of orgasm accompanied by a crescendo of the note into something like a faint version of a pheasant's crowing chorus. This was repeated over a dozen times without changing position and almost without interruption; once he began feeding in the middle without breaking off display. When disturbed he flew off with a peculiar slow owlish flight.

I have heard a flock settling to roost make a clamour of screaming and piping by no means musical sounds, very different from the typical *pee-wit* call. In Galloway in late July several birds bathing together kept darting at and chasing one another with an insistent prolonged *deeee* cry.

The nest-scrape on more or less open ground is lined with grass or other stems, and four eggs are normally laid in late March or April; there is only one brood but lost clutches are replaced as late as May or even June. They are hatched by both parents in from 24 to as much as 31 days, and the young, although running at once, take over a month to fly.

Lapwings are, thanks to protection, fairly long-lived birds; ringing returns of the British Trust for Ornithology show many cases of their surviving six, seven or eight years, and occasionally a good deal longer.

The long delicate curved crest begins to grow in the young bird and adorns lapwings irrespective of age or sex thereafter, but is laid back invisibly in flight. The male in summer has the crown, some stripes on the face and the great rounded "breastplate" black, with white cheeks, collar and eyestripe, the mantle is bronzy green with rich sheens and iridescences, and some pinkish and buffish patches. The tail is white at the base, black at the tip and there are striking bright buff tail-coverts. The male in winter, and the female and young are basically similar but with rather less and duller black and coloured markings, and more brown and white. The bill is black and the legs fleshy. The length is about 12 inches, the straight pointed bill about one inch and the tarsus about two. The weight is from 7–10 ozs, or about 200–280 grammes.

CORNCRAKE

Corncrakes, like cuckoos, are familiar voices to many people who have never seen them, and would not recognise them if they did. That persistent double rasping "arp-arp", continuing for hours on end by day and night sooner or later forces itself on the attention of the least observant, wherever it is heard. That it is not now half as widely familiar as it was fifty years ago is solely because corncrakes have ceased to breed in many of the more accessible parts of the country where they used to be common. This decline has aroused much interest, and in 1938–39 a British Trust for Ornithology inquiry was organised by C. A. Norris to assemble the full facts and attempt to throw more light on the problem.

The results showed that in England and Wales south and east of a line drawn from the mouth of the Tees roughly through Bradford, Stoke-on-Trent and Builth down to Carmarthen Bay the corncrake (see Pl. 17, p. 84) had become a scarce breeding bird, few of the county reports showing more than an odd nest or two, although handfuls of breeding pairs survived in limited areas of Nottinghamshire, Worcestershire and elsewhere. Over the rest of the mainland of Great Britain, while there had been a considerable decrease, which was in full swing at the time of the inquiry, it was still possible to find half-a-dozen pairs here and a dozen there in the most favourable areas, such as the Pennines, the Lakes, Northumberland, North Wales, Ayrshire, and Lanarkshire, where one area quite near Glasgow reported about twelve birds in six square miles. In many parts of the Highlands the position was similar, but in Kintyre and Bute, Orkney, Shetland and the Outer Hebrides, and most of the Inner Hebrides except Skye and Mull the density was much greater, the small island of Canna alone having an estimated hundred pairs in 1938. It was interesting to find that except for the narrow and remote peninsula of Kintyre the corncrake remained numerous only on the islands, and even on the islands three of the largest and nearest to the mainland (Skye, Mull and Arran) had felt a decrease. In Ireland also, which has long been known as a great country for corncrakes, a decrease had occurred over the whole of Northern Ireland except western Tyrone and Fermanagh, and much the greater part of the South, although by English standards the corncrake remained a common bird. For instance in Co. Kilkenny, despite a reported decrease, there were still estimated to be as many as 15 pairs per 100 acres in one area, and in Co. Meath about 1 pair per 100 acres. In Tipperary, which complained of a one-third decrease in ten years it was still possible to find 7 nests in one ten-acre field, and in Co. Clare up to 100 birds could be counted in 1939 within a three-mile radius of one place.

The corncrake breeds across Europe and western Asia up to the fringes of the Arctic and southward to Turkestan, North Persia, Asia Minor and the north shores of the Mediterranean westward to the Pyrenees. All the reports which it was possible to obtain from these areas prior to the outbreak of war pointed to a similar decline in numbers, particularly in Sweden, where it had disappeared from large areas in which it was formerly common, and in Finland, Denmark, Holland, Belgium and France. Only in Germany and Hungary was there no clearly recognized decline.

One of the interesting points which emerged from the inquiry was the gradual spread of the decline. Information on this point could of course not be very exact, as corncrakes fluctuate widely from season to season without regard to the long-term trend, and where observations had not been kept at the time it was often impossible to check the past history. Making allowance for these difficulties it is still significant that decrease was noted as long ago as about 1850 in Essex, 1875 in Middlesex, 1885 in Oxfordshire, 1886 in Norfolk, 1890 in Lincolnshire and Cambridgeshire, 1895 in Kent, Berks and Bucks, 1897 in Bedfordshire and before the end of the 19th century in Worcestershire and Dorset, while in Devon, Somerset, Wiltshire, Hampshire, Sussex and Surrey, Hertfordshire, Warwickshire, Rutland and South Wales, it only became apparent between 1900 and 1910. After 1910, but before 1921 the decline struck Gloucestershire, Monmouth, Shropshire, Staffordshire, Herefordshire, Leicestershire, Derbyshire, Nottinghamshire, Yorkshire and all the remaining English and Welsh counties to the west and north which had not been previously affected, or (in the case of Lancashire) had recovered from the first impact. In Scotland the situation is more obscure but signs of decline were noted in Wester Ross in 1888, in Inverness in 1895 and in Nairn and Moray about 1900, the latest areas affected being Dunbarton, Mull and Arran after 1930. In Ireland the decline began about 1902 in the Dublin area and spread to the east central counties by about 1918, but only since then has it become substantial and spread over most of the island. Almost all the reported decreases on the Continent date from after 1920.

It is clear then that the decline of the corncrake has not been a sudden and capricious setback but a progressive withdrawal over no less than a century, the one extraordinary feature being that in the British Isles the withdrawal has been steadily towards the periphery of the breeding range. To see this withdrawal in perspective it is necessary to bear in mind, as I pointed out in *Birds in England*, that there is every reason to believe that during the eighteenth century and earlier the corncrake's status in the British Isles was much more like what it is now than what it was a hundred years ago. At some time between about 1800 and 1850 corncrakes must have greatly increased in numbers and expanded their breeding range. They apparently never became common in Norfolk or Essex, and the period of real plenty in much of south-eastern England may well not have lasted

PLATE 17

Eric Hosking

Sitting corncrake in June; Orkney

PLATE 18

Robert Atkinson

Common partridge on nest exposed by burning vegetation

more than about fifty years. That would be quite long enough to convince people that corncrakes in the meadows on a summer day were part of the unchanging order of things, and that the mystery of the corncrake is why it has declined since 1850 rather than why it suddenly increased and extended its range for a few decades before that.

Undoubtedly the corncrake as a bird requiring fairly dry ground and tall herbage throughout the summer is very awkwardly adapted to live with modern agriculture under any system of farming, and it is highly doubtful whether corncrakes could achieve the minimum essential nesting success except in small and precarious patches of the great tract of southern and eastern England and Wales which they have virtually abandoned. That, however, does not amount to proof that the decline of the corncrake has in fact been caused by modern agricultural methods and machinery, although frustrated efforts at breeding may well have contributed to the withdrawal from certain areas. We really do not know enough about the conditions of breeding success for the species to trace causes. The loud obsessional and apparently unnecessary call suggests (since it can hardly persist without some biological value) that local concentrations in close neighbourhood may be important for successful breeding, and this suspicion is supported by the frequency with which corncrakes are found either in a cluster or not at all. The increasing evidence of polygamous breeding habits may give a clue to the explanation. At least we cannot rule out the possibility that some factor inherent in the corncrake's way of life, over which we have no control and of which we have still no knowledge, may be responsible. For instance, the erection of a network of overhead telegraph wires in north-west Europe historically coincided with the corncrake's decline. We know that many corncrakes were killed on certain passage routes by hitting the wires but we have no knowledge whether the scale of this mortality might have been sufficient to cause a decline in breeding numbers or whether its incidence might have been less upon birds travelling towards the north-west fringe of the range. Some more detailed study seems advisable, since we have no grounds for assuming that the decline is over, nor can we be sure that it will be checked in time to prevent the eventual extinction of the species in the British Isles.

Like quail, corncrakes make a dramatic change after a summer spent running about in a few acres of herbage by suddenly flying off to Africa, but with corncrakes the contrast is even more remarkable since they often give the impression of being practically unable as well as unwilling to fly, and yet some reach the Cape every winter. While birds flushed from cover in summer flutter with legs dangling for a short distance, at migration time they fly quite strongly, although usually low. They do not flock, but occasionally a number become bunched together at the same spot on passage. A migrant marked at Skokholm off Pembrokeshire in April has been recovered in Tipperary two months later, and one ringed in Ireland in July was found in Sussex in August, thus confirming that Irish birds pass through the southern parts of Great Britain on their journeys to and from France. Most corncrakes arrive between mid-April and mid-May and leave between late August and the beginning of November, but a few stay the winter.

In spite of its common name the corncrake is not often found craking in corn or other crops in Britain, the great majority preferring grass, and sometimes clover or rushes, in which they move about with ease and secrecy, thrusting their streamlined bodies through the thickest herbage and stopping often to crane their necks and look out for danger, which they skilfully evade. It is difficult to get a view of a corncrake except on passage, when they rise fairly readily: in Ireland many birds have been called up by artificially imitating the croaking challenge with bones suitably notched. So far as is known insects make the bulk of the food; slugs and snails, earthworms, and a certain amount of leaves and seeds of plants are also eaten.

The nest is of dead grass, in thick herbage, and usually contains 8–12 eggs, pale greenish or brownish with bold reddish brown and paler ashy blotches. They may be laid at any time from May to August but usually in late May or early June, and there is normally only one brood.

Corncrakes are about 10½ inches long, and weigh 6–7 ozs., or about twice as much as a quail. The body is rather plump but well streamlined towards the short pointed tail. The bill, rather large for a rail, is brown, and the strong longish legs are pale flesh colour. In flight the bright chestnut wings show up against the yellower body and greyish head and neck. As in other crakes the plumage is fairly well spotted, and the flanks striped protectively.

COMMON PARTRIDGE

Partridges are of all our birds the most content to stay on the earth, and moreover to stay on that part of the earth where they first saw the light. Strong on the wing as they are, they seem rarely to fly farther than across a few fields, or higher than they must in order to clear some ground obstacle. They are non-migratory, and even their weather movements are very limited in Britain. A partridge in a pear-tree is a sight less improbable in rhyme than in the field, although I did once see a young bird of the year in Surrey fly up to an oak tree and balance for some time most precariously on a twig until it lost its foothold in trying to look round at me and had to fly down again.

Farming enormously expands the area of suitable feeding grounds for partridges by keeping so many fields under plough and crops or under short grass, but partridges, like many other birds, are increasingly troubled by the inconsiderate tendencies of farmers to mess about with their farms in spring and early summer, when complete quiet is desirable, and to seek to tidy up and bring into cultivation all sorts of rough fringes and waste patches which their forefathers were content to leave untouched from generation to generation.

No form or degree of cultivation seems too much for partridges, nor do they show any clear preference between arable and grass fields, provided that the necessary minimum of rough herbage and low bushes remains somewhere accessible. In search of these conditions breeding partridges will resort even to large gardens, roadsides, railway embankments, airfields, and other enclosed places close to suitable open country. I have flushed a pair inside the enclosure of Stonehenge, and they sometimes occur on commons thickly grown up with tall bushes, and occasionally even in woods. On the Yorkshire moors, and no doubt elsewhere, partridges may be found well up in the heather overlapping with red grouse. On the marshes they overlap with waders such as redshanks. While they avoid built-up areas I saw a 1949 covey just reared in the County of London within six miles of Hyde Park Corner, and another even nearer site was frequented at least up to very recently.

Partridges are lowland and upland but not highland birds. I have flushed a covey in December after they had settled to roost at dusk on Cleeve Cloud in the Cotswolds at over 1,000 ft., and the highest elevation at which I have noted any was at over 1,100 ft. on the Shropshire-Montgomery border.

There are few regions of the British Isles where the partridge is unknown, except in the outer Scottish islands and parts of the Highlands. Nevertheless the great bulk of our partridge population is massed in the eastern parts of Great Britain, from Cromarty Firth southwards. Along the Channel coast fair numbers are met as far west as the South Hams of Devon, but partridges are distinctly sparse over much of Devon and Cornwall and in Wales. In Ireland I have never seen one, although they occur in all parts. On the Continent, where several separate races are distinguished, partridges breed as far north as Trondhjem and Archangel, and eastwards into Asia, while the southward limit is marked by North Persia, Asia Minor and the Mediterranean. They have been introduced successfully in many parts of Canada and the United States.

In really suitable areas, and especially where they are actively preserved as game, partridges will in good years attain remarkable population densities. On one Norfolk estate, Witchingham, the numbers actually shot averaged very nearly 1 per acre in one season, and there were almost as many survivors. There are a number of records for estates of several hundred acres in eastern and midland counties which carried an average of between one and two partridges per acre at the end of the breeding season. In such areas the partridge is at times the most numerous species of bird, and taking the average weight at 13 ozs. the most favoured areas will have to support perhaps as much as a ton of partridges to every 2000 acres. Some species run to even greater densities, while others are even heavier, but few exercise so great a combined pressure over such wide areas upon the means of subsistence. It is due to the high fertility of our land and to the flair of partridges for selecting as food so many plants regarded as weeds by farmers that this great living burden can be carried without creating any significant difficulties either for partridges or for people. Probably the covey-forming habit is the secret of this success. If partridges were given, like rooks and woodpigeons, to massing in hundreds on cultivated fields, even if their total numbers and diet remained the same, it would hardly be long before any actual or expected crop failure began to be laid at their door in emotional speeches and letters to the Press describing how the fields were brown with these voracious pests. But as partridges have the wit to be unobtrusive in colour and manners, and to limit their bands normally to a couple of dozen birds or less, and as they pay toll to sportsmen and are in return protected by them against other potential destroyers, any damage which they may do to crops is overlooked instead of being picked out, harped upon and exaggerated.

Partridges have many enemies. Foxes often spring upon and kill sitting hens, and swallow the eggs whole. Badgers sometimes do much the same (one having been known to destroy eighteen partridge nests in a week) and so do dogs, cats, stoats and weasels, while rats and rooks account for quite a number of eggs. Farm workers and livestock often unintentionally destroy the nests, and sometimes the birds themselves are killed or mauled by farm machinery. In wet and cold summers the recently hatched young die in great numbers from wet or exposure, and a sunny June and July is essential to a good partridge year. Several diseases also take toll, not to mention poachers and a variety of other causes. A. D. Middleton, who has made a thorough study of the species, drew up the following annual balance sheet to show how an average population of 100 partridges, divided equally between the sexes, reproduces itself and drops back to approximately its initial strength next season:—

1.	Total nests	50	
2.	Total eggs laid, taking av. clutch 14.6 eggs	730	
3.	Loss of nests 22 per cent, reducing (2) to	579	
4.	Eggs not hatched 7 per cent reducing (3) to	538	chicks hatched
5.	Chick mortality 52 per cent leaving	259	young in Aug.
6.	Add original parent birds	100	
7.	Total stock before shooting (5) plus (6)	359	
8.	Stock less 55 per cent shot	162	
9.	Stock less 40 per cent "winter wastage"	97	

These figures are of course rough averages, based on figures from a number of estates in different parts of the country and in some cases over a period of years, and thus seeking to combine a wide range of different experience. What emerges is that partridges reproduce on a scale sufficient to ensure an enormous rate of increase if there were not in fact an enormous rate of loss. Even under vigilant game preservation only about two-thirds of the eggs laid hatch out into young partridges, and normally over half of these, or in bad years many more, have perished by August, leaving a new generation at that stage equivalent to about one-third of the eggs laid, but two-and-a-half times the strength of the parent stock. Thus of every seven birds on the ground at the beginning of August about five must either perish or move to fill gaps elsewhere if the population is to remain steady. Middleton concludes that many estate owners, in their eagerness to maintain the maxiumum population, take care to leave a bigger stock after shooting than will in practice be able to stay and breed after allowing for natural losses, with the result that a surplus population must leave the estate each spring in search of other homes. It is no doubt quite normal for many species to produce, in those areas where they can breed most successfully, a considerable surplus who must face trying their fortune elsewhere, since if this were not so there would be no source to feed extensions of range by colonising new ground, or even to maintain numbers in the inevitable deficit areas which in some or all seasons suffer too heavy losses to maintain their own strength without some reinforcement from outside. The process is simply more obvious in the case of partridges owing to the sporting interest in their numbers, the large clutches of eggs laid, the absence of second broods, the relative ease with which they can be counted at all times of year, and the very limited distances which they normally cover.

Partridges begin pairing early—I have seen a couple in Lincolnshire as early as the 2nd January in a mild winter. At this time they become very noisy and active. I have seen two cocks running up and down as fast as they could go, sometimes one in pursuit and sometimes the other, keeping close to a wire fence and constantly reversing direction after chasing for a dozen to a hundred feet, either by the fugitive bird dodging and doubling, or by it taking turn as pursuer. Breaking into a fight they stopped, faced each other, bowed, and struck each other's faces with their bills, varying the performance by springing up about a foot into the air with whirring wings and striking at one another with their feet like game-cocks. One drove the other through the fence and they went at it again. Now the loser took flight and sped across the field, the other after being taken by surprise following and catching up. When they settled one bird, apparently the loser, stood up in a fighting attitude while the other crouched almost hidden in the grass, even when the first bird strutted up to him. Then a third much smaller bird joined, and all three began racing up and down the field in Indian file, the whole performance being most formal and the movements beautifully synchronised, both in fighting and in chasing. In late summer I have seen very undersized young birds in a Yorkshire covey repeatedly leaping up in friendly sparring.

Once in the middle of March on the Chilterns I flushed from near a hedge a hen partridge who flew out with the right wing stiffly half-folded and beating ineffectively for about forty yards, losing height gradually until she dropped to the ground and began running away. The cock, who followed after an interval, settled ahead of her but let her catch up and ran by her side. The wing was not trailed after alighting, and there was no trace of an exceptionally early nest where she rose. I was left guessing whether to regard this as an unseasonable exhibition of "injury-feigning", or whether the bird genuinely had wing injury which did not show until she flew. When flushed with chicks it is common for the hen, and at times both parents, to give a frenzied exhibition of injury-feigning.

Partridges are often found on a roadway. Once when I flushed a covey in a deep Devon lane one bird failed to clear the very steep overhanging bank and dropping back began running in front of me, beginning with a sprint of at least ten miles an hour but wearying rapidly after about fifty yards, when it was very relieved to find a gate through which it could double into a field and out of sight.

Partridges fly fast, having been timed at from 25 to 35 miles an hour airspeed in normal flight, and up to more than 40 when pressed. The whirring wings are kept below the bodyline, and there are frequent brief glides, but it appears fairly well established that the flight is as tiring as it looks, and can only be sustained over distances which to many small birds would be insignificant.

Among the many things which we do not know about this common bird is exactly how it roosts, although the word "jug" has long had a special meaning as describing a covey of partridges settled down to roost in a bunch on the ground. Many writers have followed one another in suggesting that the birds roost in a circle, some say facing inwards and others outwards, while it has also been suggested that all probably face the wind when there is any. Christopher M. Swaine has described seeing a covey of about fifteen on a windless, bright moonlight late summer night in North Wales resting in an arc with all heads facing outwards except for one standing alone between the two ends of the arc. This bird faced inwards. (See Pl. XIII, p. 80).

The hoarse creaky *turr-wit* note has great carrying power, and is heard as a challenge wherever cock partridges are in possession, while the indignant clucking of a flushed covey is also a very familiar country sound.

The nest (see Pl. 18, p. 85) is a scrape in the ground lined with dry grass and dead leaves, sometimes sheltered only by low rough herbage, but more often protected by a hedgerow, bush, or shrub, or by tall growing crops. The pale olive-brownish eggs vary greatly in number, a single hen laying anything from eight to 23, and it frequently happens that two or more hens lay in one nest, producing as many as 40 eggs in extreme cases. The average appears to be nearly 15. They are laid in April or early May, and take 23–25 days to hatch. The hen will sit extremely close, and I was once compelled to give up the attempt to persuade one to leave by methods short of lifting her off by force, after trying for ten minutes within arm's length. While laying, and sometimes later, the hen protects the eggs on leaving them by covering them with light material such as grass stems. Although they leave the nest at once the young are not able to fly for about sixteen days, during which they are extremely vulnerable to many sad fates.

Partridges are very plump birds with relatively small rounded heads and tiny bills hardly longer than a greenfinch's. They are about twelve inches long and stand rather upright on stout medium-long legs. The brown and grey plumage has bewilderingly complicated markings, the general effect of the upper surface being a warm reddish-brown, the flight-feathers dull brown, the breast fairly pure grey and the lower breast in the male marked with a very prominent dark chestnut horseshoe-shaped patch. The tail is chestnut and the flanks have bold chestnut bars.

RED-LEGGED PARTRIDGE

(Pl. XIVa, p. 81)

Man is directly responsible for the existence of the red-leg as a British breeding bird, and it is not certain that the species has ever reached this country without human aid, although occasional immigration from the Continent has been suspected. It was Charles II, that great bird-lover and friend of scientific experiment, who first introduced this species at Windsor in 1673, and although these birds did not survive his good judgment is vindicated by the fact that red-legs are now flourishing within sight of Windsor Castle. About 1770 large numbers of eggs were brought over from the Continent and hatched

PLATE 19

Eric Hosking

Young cuckoo ejecting egg from tree-pipit's nest

Eric Hosking

Long-tailed tit feeding young

under fowls in east Suffolk, and further introductions followed, the history and results of which would repay more study than they have yet had.

As usual, the realisation that the introduction had been a success was quickly followed by a widespread suspicion that it had been a disaster, and by an abortive attempt to exterminate the newcomer. A. E. Knox wrote just a hundred years ago "I rejoice to say that this species is not indigenous to Sussex. Many a Norfolk and Suffolk sportsman has to suffer for the sins of his fathers, who unwittingly introduced this foreign plague into their ancestral domains. Some portions of these counties are fortunately exempt from them, while in others they have increased to such a degree as to expel the old English or cinereous partridge, and being excessively wild and difficult to flush, they run before the dogs for miles and severely test the patience and temper of the sportsman". This rejoicing was premature, for, although Knox was able to record the failure of an introduction into Sussex at Kirdford in 1841, red-legs were soon to establish themselves firmly, by a mixture of natural expansion and local introduction, right over south-east England. In 1865 they began to spill over from Hertfordshire into north Middlesex, and within about twenty years were firmly established over most of suitable areas between the Humber and the Thames, and inland to Cambridgeshire and the East Midlands. The range has since continued to extend westwards as far as east Devon and Somerset, into the Severn Valley, and even as far as North Wales and Yorkshire. Attempts to introduce them into Scotland and Ireland have failed. They are only at home on certain soils, especially those which are sandy, alluvial or chalky, and are missing from clayey and other unsuitable areas. Apart from this artificial and isolated northern outpost in Britain red-legs are no longer resident in north-western Europe, but are found from western Switzerland and central France, south to Central Italy and Corsica, while allied forms inhabit Spain and Portugal and the Atlantic Isles. Other similar species of the same genus are found from the Alps and North Africa eastwards to the Himalayas.

Both geographically and ecologically the red-leg has occupied that part of the common partridge's range which is most favourable to it, without displacing the common partridge, so that the common bird is often found without the red-leg but the red-leg usually side by side with the common. Where the red-leg is found alone it is usually in habitats too extreme to be attractive to the common partridge, such as big tracts of shingle (even on the seashore) marshes and almost bare rabbit-warrens. The fears and assertions about the driving out of common partridges by red-legs have definitely not been fulfilled.

Although red-legs are introduced birds in England I am sure that they sometimes perform some kind of migratory movements. In an area on the middle Thames which I was keeping under the closest observation during the early months of 1925 I found not a trace of red-legs until February, when a pair came in. The next day I saw two more pairs close by, and the following day three at a third locality. On the 26th and 27th and at intervals during the following three weeks I made a very careful search of the same ground and the birds were no more to be found. I believe they had been passing through on the way to the high Chilterns where I found them in March up to 700 ft.; they occur at similar elevations on the Sussex and Hampshire Downs. Several of the well-recognized seasonal movements of other species to breeding quarters were evident at the same time in the same area.

Red-legs are recorded as perching fairly frequently on fences or even trees, but I have not myself observed this. On the ground they take alarm at much longer range than common partridges, and lose no time in running away very fast and very far, not flying if it can be avoided, and then taking wing from as distant a point as possible, usually to pitch again after surmounting some obstacle or crossing to the next suitable field. Red-legs are heavy birds, the males weighing upwards of 1 lb., and they fly strongly and fast.

When on the alert the outline looks more curved and less erect than in the common partridge and the head smaller in proportion; they are larger, darker slatier birds, and in reasonably good view the more contrasting effect of the white cheeks and throat against their black border, and the boldly barred black, white, and chestnut flanks, as well as the bright red bill and legs make the species unmistakable. Birds flying or running more or less directly away are often hard to identify with certainty on the spur of the moment, and the reddish tail increases their similarity to the common species.

By far the best means of knowing whether red-legs are about is by listening for the far-carrying, unmistakable, monotonous piston-like "song" with its mechanical rise and fall, which has always sounded to me like *corictishar*, *corictishar*, *corictishar*, with the accent on the last syllable, often repeated many times without pause. This note, which might not be guessed to come from a bird at all, is uttered very persistently at all times of day through most of the year.

The food is similar to the common partridge's, but insects seem to be even less eaten, and some recent samples show large proportions of sugar-beet and cereals.

The nest is also similar, and is similarly sited, apart from aberrations. I have seen one with 17 eggs almost awash with salt water on the saltings behind Scolt Head, which, like Blakeney and Dungeness is a favourite breeding-place. Eggs are laid in bewildering profusion, from 7 to 28 being deposited by a single hen in a single nest, and it has even been suspected that the hen sometimes lays two separate clutches one of which is incubated and tended by herself and the other by the cock. At times a red-leg will lay in a common partridge's nest. As a schoolboy I was astonished after putting a red-leg off her nest full of eggs to discover a complete clutch of common partridge's underneath making a combined total of 33. The eggs hatch in 23-24 days and are dirty cream colour freckled with brown and more or less blotched with dark brown. I have seen "injury-feigning" on surprising one with chicks.

QUAIL

(Pl. XIVb, p. 81)

To go in quest of quail is one of the most baffling of ornithological enterprises, and I must admit at the outset that I have not actually succeeded in watching quail for more than about thirty seconds in the last thirty years, that being about the time it took two flushed after a long search with a dog to get down into impregnable cover over the first hedge. Before that, on 25 November 1917, I was lucky enough to put up twice in succession a flock of about fifty of these birds, no doubt on passage, near Cissbury Hill on the Sussex Downs, and noted that they looked like partridges reduced to the size of starlings, and were fairly shy. Such flocks, formed out of several families, are exceptional to-day in the British Isles, and as far as is known most of the birds which now breed with us depart about October, although wintering used to be frequent, and in Ireland regular.

The quail is alone among our game-birds in being a long-distance migrant. It has an enormously wide distribution, one race or another spanning the whole of Europe and Asia from Ireland to Japan, from not far below the Arctic Circle down to North Africa, Asia Minor,

Persia and North-west India, while wintering birds occupy a long strip from the Mediterranean down to tropical Africa and across Arabia, India and Burma to the Pacific coasts from Siam up to South China. Another race occupies southern Africa. From April to mid-May the enormous numbers of quail migrating through the Mediterranean region have been massacred in bulk for food since the beginning of historical times. The late Sir Hugh Gladstone calculated on the basis of data given in the Book of Numbers that about the year B.C. 1580 the Children of Israel killed within thirty-six hours in April upwards of 9 million quail at the place afterwards called Kibroth-Hattaavah. Be this as it may, modern records show that the numbers taken annually during the three weeks' spring migration on the Island of Capri fell by degrees from 150,000 in a good year about a century ago to 56,000 about half-a-century back and 30–40,000 by about 1904. On the Naples coast 100,000 were reported to have been taken in one day within four or five miles in 1867. In South Italy the slaughter seems to have been even greater. The Prefecture of Messina reported the three largest companies in the trade as having exported a total of 536,500 quail in ten days on spring migration, and in 1895 the Port of Marseilles cleared 8 million quail, mostly from Brindisi and Messina.

In the autumn of 1890 620,426 were recorded as coming in to Udine market and in October 1889 423,000 passed the gates of Brescia. A great diminution not unnaturally followed. The destruction of nests in Italy was also immense.

In Egypt the trade seems to have developed rather later, with an export from Alexandria to Europe of 300,000 living quail in 1885, rising to 2 million in 1897 and falling back to 1,275,490 in 1898, of which the great majority were consigned to France, although almost all were said to be eventually destined to be eaten in England. In 1920 the annual export from Egypt was given as over 3 millions. In the spring of 1924 on an island off Italy six guns killed 3,806 quail in a month. Whatever discount may be placed on these figures, which were not scientifically checked, it is unquestionable that the toll levied on the quail of western and northern Europe must have amounted to a severe strain on even an abundant and prolific species. It should be borne in mind that this toll of birds which had survived the winter and were ready to breed is more serious than a considerably heavier destruction in the autumn, and that quail in many areas had already had to run the gauntlet of heavy fire from sportsmen in their home

PLATE XV.

Eric Hosking

(*a*) Pair of jays

Arthur Brook

(*b*) Magpie with young

Eric Hosking

Cock blackcap, aggressive display

countries and on the southward journey as well. While we have no definite knowledge of the destination of the birds slaughtered on their way north at Capri there is reason to suspect that some of them were native to the British Isles. One ringed farther north, at Brescia in Lombardy on 29 May 1933 was killed by a mowing-machine in Sweden, near Stockholm, twelve weeks later.

In view of the evidence of huge destruction on spring passage and of declining bags it would naturally be expected that the breeding stock in the areas served by these migration routes would have declined too. As quail are so difficult to observe and their numbers in any case fluctuate considerably from year to year the record is not altogether plain, and some have questioned whether any genuine diminution has occurred, but the evidence to that effect is too strong.

Gilbert White knew quail at Selborne as rare and apparently irregular summer migrants, but instead of claiming them as he would now as remarkable, he apologetically attributes their local scarcity to the fact that the area is too enclosed for their liking. He also said that they "crowd to our southern coast and are often killed in numbers by people that go there on purpose". There was a time when quail-shooting was a recognized form of sport in Britain, but it has died out in the past century. It has often been suggested that improved cultivation has driven out the quail, but nests in recent years have too often been in highly cultivated areas, and actually among crops, for this view to be accepted. The natural and obvious conclusion to be drawn from all the available facts is that the improved market for quail and the means of destroying more of them more profitably during the nineteenth century in the Mediterranean gradually undermined our breeding stock. The only inconsistent evidence is the often-repeated statement that the Irish stock which largely disappeared about a century ago was mainly resident, which if correct would point to some factor affecting the birds in the British Isles as responsible. In the absence of fuller evidence, however, it seems more reasonable to suppose that the Irish quail shared the migratory habit of all the others from northern latitudes, but that winter stragglers were sufficiently common and often enough shot in Ireland to create an exaggerated impression of general year-round residence.

As since 1939 the European market has been much disorganised and some of the coasts on which they halt have been heavily mined against invasion, it might be supposed that the past decade would have been more favourable to decline in persecution and therefore to some recovery in quail numbers. Recent experience is quite consistent with this view. While some have been reported in most years from such favoured areas as the Cotswolds or the Berkshire Downs breeding in 1943–44 was evidently on an appreciably increased scale. In 1943 eight or nine pairs probably nested on Fair Isle, and in 1944 six birds were heard calling in quite a limited area in East Yorkshire, and breeding was recorded in two other localities, while in one Wiltshire area out of several pairs present four probably hatched off, in addition to a fair number of more scattered records from well over a dozen other counties in England and half-a-dozen in Ireland. R. E. Moreau has estimated that in 1948 at least twenty different quail were calling during the summer in a strip of country about nine miles long by one-and-a-quarter broad on the western foothills of the southern Chilterns in Oxfordshire. Reports of increase also come from the Cotswolds, and making every allowance for annual fluctuations and for possible improvements in recording it seems clear that some revival in quail numbers is occurring. The maintenance of this revival may well depend on the maintenance of a depression in the quail trade of the Mediterranean countries. It is to be hoped that a world which talks so much of European co-operation will be able to achieve co-operation in not decimating our quails. It must, however, in fairness be admitted that British appetites have in the past too often been behind the slaughter of quails in Mediterranean lands.

In the breeding season quail are usually located by hearing the triple *wet-my-lips* call-note either by day or at night, for they are active after dark. But Mr. T. J. Willcocks who kindly and skilfully showed me some quail which had bred on his farm in Cornwall told me that neither he nor his farm worker was aware of having any quail on the farm that season until late in August when a whole brood, were found on reaping a barley field not far from his farmhouse. Both were familiar with the call and were on the lookout as quail had nested previously on the farm, and Mr. Willcocks is about as good an observer as any in England. It cannot therefore be assumed that if there is no calling quail are not present and breeding, and it is possible that a good deal of calling may be by non-breeding birds. Cases are on record of the call being uttered in flight. The low note uttered on being flushed sounded to me like a somewhat purring partridge call.

Although a bevy of ten had been put up a fortnight earlier the dog was only able to find two for us. They rose only a few feet up, looking very small, round, and partridge-like with very fast wingbeats and general earthy-brown colouring except for conspicuous striations of buffish and very dark brown on the head and apparently also the mantle. We immediately followed them up to close marks but were unable to flush them again. They came up from a small uncultivated patch of dense herbage, at the bottom of the 16-acre barley field where the brood were found. A previous nest on the same farm was in a 20-acre barley field, the altitude being about 150 ft. up and the country gently sloping with low hedgebanks giving it a definitely enclosed character.

Quail have long been famous for their bellicose natures. In most respects they behave, as they look, like miniature partridges. They lay from seven to twelve eggs in a poorly lined scrape in the ground, the shells being so heavily blotched or spotted with chocolate as to resemble a miniature red grouse egg rather than a partridge's. They may be laid any time from May to September and hatch in from 16 to 21 days, after which the young are flightless for a period, but can run. Quail are only about 7 inches long and their weight averages 3½ ozs., or roughly a quarter of a partridge's. The male can be distinguished from the female by an anchor-shaped dark marking which begins below the bill and, after forking, curves across both sides of the white chin.

CHAPTER 5

GARDEN, ORCHARD AND HEDGEROW BIRDS

Of the world's total land area of 35,700 million acres only some 3–4,000 million acres are actually cultivated, and of this total of the world's farmland roughly 1 per cent is in Great Britain. We have no knowledge of the total number of gardens on earth, or how much of the earth they cover, but we may safely guess that their extent is diminutive compared with farmlands and that a much larger share of it is to be found in England, Scotland, Wales and Ireland—but above all in England. From the air it is remarkable to see how much of England has actually been made into gardens, and how much of it consists of landscapes designed by men with gardeners' eyes.

English farms with their hedgerows and banks, their shade trees and shelter-belts and orchards, are ecologically often more akin to gardens than to the farms of most of the outside world. Even English towns, except the too numerous wens of nineteenth-century industrialism, show the influence of gardeners matching or exceeding that of architects and builders, and the more modern the town the more the gardener pervades it. As the raw new suburbs march outwards across Middlesex, and Warwickshire, and the West Riding, lawns and shrubs and flower-beds struggle to enfold them. In the blitzed centres of large towns the dust and wreckage begins to form itself into gardens, sometimes like minature wildernesses, sometimes formal and well-tended.

These clusters of gardens are as yet too little appreciated, partly because recent building has often raced ahead of the gardener so that too many gardens are too new to be seen to mature advantage, but also because the immense revival of British gardening has not yet made itself fully felt. Many town and suburban gardens are still in the immature and formative stages, out of which most of our great parks

C. Douglas Deane

Tree-creeper at roost, under drifted snow

PLATE XVIII.

C. W. Teager

Blue tits at coconut

and country house gardens only emerged after the death of those who first planted them. While, with some skill and some luck, flowers, lawns and even rockeries can be fairly quickly contrived, it takes years for the garden to assume its true character, and for the trees and shrubs to grow until they can provide good harbourage for birds. So during the past thirty years hundreds of thousands of gardens have been laid down in Britain which will in course of time support and sustain many more birds that they yet hold.

Of all habitats gardens are normally the richest in bird life, provided that they are not too small, too blighted by smoke and fumes, or too infested by cats. A British Institute of Public Opinion sample makes the cat population of the County of London alone well over half-a-million, and in the country as a whole there are undoubtedly several million cats—nearly 14 million according to the National Animal Registration Service. The majority of these, being housepets, range over gardens and must kill very large numbers of birds, especially young birds, and chiefly small songsters, such as robins, blackbirds, and throstles. In some areas, where they are especially thick on the ground and where food and nesting sites are in any case limited, cats reduce and keep down the numbers of songbirds. Even more unfortunate is their influence in depriving us of the pleasure of increasingly tame and approachable garden birds, by the heavy toll they take of those birds rash enough not to remain perpetually wary. Many well-meaning people are guilty of the greatest cruelty in keeping cats and at the same time trying to tame their garden birds, which in responding fall easy victims to one of the more unpleasant and painful forms of death. It should be obvious that those who take pleasure in taming wild birds should not keep or encourage cats, and vice versa. But for the fact that they are swarming with cats our town gardens and parks would undoubtedly shelter far more and, above all, far tamer songsters.

Gardens are of many sorts, some attractive to few birds and others to many. At the poorest end of the scale are town roofgardens or backyards with a few shrubs and plants, perhaps even in tubs and window-boxes. These, and their rural counterparts support little beyond house-sparrows, and make merely a slight addition to the amenities of the buildings they adjoin. Then there are plots of some small fraction of an acre with perhaps two or three trees and some shrubs and creepers. When several of these are grouped together, even in the midst of a town, blackbirds and blue and great tits begin to find a foothold, while in slightly larger and richer gardens throstles, robins and dunnocks become frequent. Large gardens in suburbs or in open towns, and quite small gardens in suitable country, begin to fill up with chaffinches, and with wrens where suitable cover exists, together with coal-tits where there are conifers, greenfinches where there are suitable shrubberies, and pied wagtails and spotted flycatchers where there are creepers or other suitable nesting sites. Starlings, although so urban, only become prominent as garden birds where the gardens are fairly spacious, or form a large block together. Jays, hawfinches, goldfinches, bullfinches, tree-creepers, nuthatches, long-tailed tits, goldcrests, willow warblers, blackcaps, lesser whitethroats, and mistle-thrushes and tawny owls become regular in gardens only where they are exceptionally large and mature, or where they adjoin a block of favourable woodland. Very locally certain other species occur in gardens—cirl-buntings, wrynecks and nightingales in the south, pied flycatchers in the west and north, and redstarts, although much less often in England than on the Continent.

Many other species often occur in gardens without being garden birds. I know a cliff-top garden in Suffolk where ringed plovers are to be seen on the lawns, a garden in Surrey where woodlarks breed, and a garden in Sussex which is haunted by grey wagtails. Rooks often nest in trees which happen to stand in gardens, and moorhens will not hesitate to settle down on ponds which are surrounded by garden. Even woodpeckers will breed not infrequently within the bounds of a garden, but the number of species which regularly prefer or even tolerate typical gardens as habitats is much more limited than the number which can be found in gardens here and there more or less by chance.

As habitats gardens vary a great deal. Some are comparable to windswept islands or brackish marshes, being adverse habitats in which owing to smoke pollution or some other cause comparatively few species of plants and insects will flourish. Others are in process of being created out of some different habitat such as field or woodland which still greatly influences them, or are dominated by buildings. Some are very formal and much tended, others like wildernesses. Some are incessantly disturbed, others are quiet sanctuaries. Many have some special feature such as conifers or berried shrubs or ivied trees, or a pool or a stream which gives them a special character. The amount of cover and the division of the open surface between lawns, flower-beds, vegetables and soft fruits and such features as nettles and docks, dung-heaps and compost, poultry runs and outhouses, espaliers and hedges greatly influence the bird life. Yet with all this diversity there are probably few gardens worth the name which are not inhabited or freely visited by great and blue tits, throstles and blackbirds, and robins and dunnocks, while chaffinches, greenfinches, wrens and spotted flycatchers are also highly characteristic of British gardens.

Akin to gardens, but often in larger combined blocks of uniform character, are urban parks, churchyards and cemeteries, allotments and nursery gardens, planted roadsides, railway banks and reservoir embankments, approaches to airfields and sewage farms, tennis courts and bowling greens, and many other patches of land which have been caught up, planted and tended in some state intermediate between field or forest on the one hand and the sterile surfaces of civilisation on the other. The birds of such garden-like habitats are distinctly similar to those of gardens themselves, but often show a tendency to include several hedgerow and heathland species such as yellowhammers, linnets, whitethroats and red-backed shrikes, or woodland species such as bullfinches, marsh tits and woodpeckers, or field birds such as partridges and skylarks, according to the type of more natural habitat which they most resemble.

Orchards also have much in common with gardens, but when at all large they are basically, like conifer plantations, areas of specially planted and controlled trees largely of exotic varieties, of one or a few species, and of uniform age and growth. Fortunately, unlike conifer plantations, they admit plenty of light and air and do not suffocate the herbage and impair the soil beneath them. Old orchards are often outstandingly rich in arboreal birds, but in recent plantings wholesale banding and spraying with insecticides, and other attentions of the fruit-grower are unfavourable to bird life. No doubt differences in types of orchard are reflected in differences in their bird life, and the study of these differences deserves more attention than it has yet had.

In England and Wales apples outnumber all other fruit trees together. In 1944 there were 17½ million apple trees on agricultural holdings of an acre or more, just over half being dessert apple trees, while of the remainder about a quarter were cider apples and three-quarters cooking apples. There were under six million plum trees, which came next in abundance, followed by two million pears, 840,000 cherries, and 350,000 nuts. Probably all these trees, unless heavily and regularly treated with insecticides, are exceptionally rich as sources of insect food, and all produce buds which are of interest to tits, bullfinches and other species. The actual fruits however vary a good deal in palatability to birds, and therefore in liability to attack, cherries being the most vulnerable tree-fruit and apples the least, which is interesting in relation to the fact that apples are also so much the most numerous.

But differences in species of fruit trees grown are not necessarily the most significant in determining the bird life of orchards. Some orchard land, for example, carries fruit trees and herbage only, while in other cases the ground beneath is cultivated for soft fruit, flowers or vegetables. In 1944 there were just under a quarter of a million acres in England and Wales with crops, fallow, or grass below the trees, while only 15,000 acres had soft fruit under trees and 18,000 acres were under soft fruit in the open, of which half were accounted for by strawberries. Nearly half the soft fruits under orchard trees were gooseberries, the next largest class being red and white currants.

The age of orchards is probably quite as important as their type in influencing bird life. Like gardens orchards are expanding and therefore a large proportion are at present immature. On the other hand there is also a fairly high proportion of ancient orchards which economically ought to have been grubbed up and replanted, but fortunately for the birds have so far been neglected. It is these old orchards which are especially rich in natural holes for breeding, roosting and foraging in severe weather. Young orchards, even after they have come into bearing, are poor in nesting sites, and modern well-tended orchards are also deficient in this way unless nesting boxes are provided. Few orchards however, form large enough blocks to stretch far beyond the foraging range of birds based on adjoining woodlands, gardens or hedges. Nearness to water and other factors also play a part.

It is open to doubt whether birds are the important agency in controlling insect pests of orchards, or on the other hand in taking toll of the national fruit crop, that they have often been supposed to be. Over the country as a whole their influence in either direction is probably slight, although in particular areas it may well be material at certain times. But the influence of orchards upon bird life is considerable and is growing with their extent. Certain species of tits, especially the blue and great, and of finches, especially the goldfinch and bullfinch find in or around orchards an exceptionally favourable breeding habitat, and over a long period the spread of orchards may be expected to strengthen such species.

In orchards the species and varieties of trees planted, their spacing and their pruning and general treatment are governed within narrow limits by economic and scientific considerations, which also determine the choice of altitudes, soils, slopes, and exposures for the siting of the orchards themselves. The conditions have to be suitable not merely for the trees to grow, but for them to bear ample fruit. The orchard acreage is accordingly massed in those parts of lowland Britain, especially in the south and west, where good soils and mild sunny climates come together.

Where trees are planted to give amenity to churchyards and cemeteries, public parks and gardens, embankments of highways, railways and reservoirs, or approaches to schools and institutions, the choice of site is usually predetermined by extraneous requirements of civilisation, and the object is not to grow a crop but to cover up some blot on the landscape or to create additional amenity. While many of the ordinary native trees and shrubs of Britain are used for these purposes there is a long and growing list of ornamental and amenity species imported from all parts of the world, or propagated by gardeners to adorn such places. The Victorian taste ran to dark evergreens, and resulted in a rash of laurel, speckled aucuba, privet, and *Rhododendron ponticum* which formed the open-air counterpart of the aspidistra. While these plants often proved able to survive in polluted atmospheres they are singularly unfavourable to most types of bird life, and their popularity has kept down the numbers of birds of many of our town parks and gardens. Yews, cypresses, hollies, box and some other evergreens, which became equally popular, are often attractive to birds, and the Wellingtonia, often planted in avenues, is a special favourite of tree-creepers, which scratch out hollows in its bark for roosting. (See Pl. XVII, p. 100).

Fortunately the modern fashion has swung in the direction of berried and flowering shrubs, most of which are far more suited to birds. Some are among the favourite food-sources of such a rare winter immigrant as the waxwing, which is coming to share with the black redstart the status of an urban rarity hardly to be found in the countryside. Bamboos, pampas grass, ivy, osiers and other shelter masses which hold plenty of food in severe weather are also invaluable in enabling delicate and specialised birds to win and keep a foothold. Among exotic shrubs the berberis, cotoneaster and pyracantha are both popular and satisfactory, while native or cultivated broom, gorse, honeysuckle, dogwood and bramble, and among trees the rowan, alder, birch, beech, hornbeam, oak and ilex all conform to the needs of amenity and bird life. This list is by no means exhaustive, and no one should be afraid to experiment with other species, so long as it is borne in mind that coarse, thick-leaved evergreens which take most of the light from the ground and which offer little food to insects or birds should be avoided. If all the aucuba and most of the privet, laurel and *Rhododendron ponticum* which at present cumber so many of our municipal shrubberies could be replaced by trees and shrubs of more suitable types a welcome increase in songsters could be expected. At the same time the main cause of the dreariness of so many of these places would be eliminated. The more our gardeners brighten up their choice of shrubs and ornamental trees the more they will be helping to attract and support bird life.

Hedgerows, together with windbreaks, avenues, shade trees and amenity clusters, are the means of carrying across the countryside some of the grace and hospitality afforded by our gardens. In England especially we have come to take our wealth of hedgerows so much for granted that we readily forget how new an invention they are, how few other regions of the world are blessed with them, and how precarious their future may be as farmers' needs and methods change. Quickset hedges can probably only flourish where the climate is mild and damp enough for their growth, but windy and changeable enough to call for their shelter, and where men are prosperous enough to let them grow and mature from generation to generation. Climate, agriculture, sport, and an eye for landscape have combined to endow us with the world's richest inheritance of hedgerows, which stretch in endless living skeins and garlands back and forth over the face of lowland Britain.

These hedgerows harbour and feed a substantial part of our bird population. There may easily be as many as 15 or even 20 miles of hedge to a square mile of country, and birds may average as much as one to every 4 or 5 yards of hedge. The blackbird, throstle, chaffinch, yellowhammer, dunnock, robin, wren and whitethroat are perhaps the most generally abundant and characteristic hedge-birds, and although all these are found freely in other habitats it is often in hedges that their greatest local strongholds are to be found. Linnets, blue and great tits, long-tailed tits, magpies, willow-wrens, lesser whitethroats and breeding partridges are among other species which depend considerably upon hedges of various sorts, although normally having their main strength in other habitats. The difference in bird life between gardens and hedges is less in the species present than in their relative numbers. The thrushes and their immediate relatives tend to be even more dominant in hedges, while tits and finches are usually less so, unless the hedges are unusually big. House-sparrows are fond of hedges in suitable patches of country, and sometimes also starlings, but both these are more general in gardens. The slightness of the difference is understandable when it is borne in mind that a field surrounded by hedges differs from a garden mainly in the larger open area in its centre and the usually more limited variety of plants grown in it. Neither of these points are very significant to most hedge-dwelling birds, although the increased elbow-room may be vital to field species.

Market gardens and allotments represent a hybrid habitat between garden and farmland, being agricultural units so miniature in scale and often so close to urban or suburban areas as to resemble gardens in many respects. In London since 1939 the emergency allotments formed in public open spaces have provided increased food resources and relative freedom from disturbance to a number of birds, including the starling, blackbird, robin, dunnock and greenfinch, while one new colonist, the goldfinch, has been enabled to establish itself sporadically on weed patches on allotments after first appearing on the rough herbage which grew up round air raid shelters, gun and rocket sites, searchlights, and other emergency works.

In the neighbourhood of Roscoff in Brittany, onions, greens, artichokes and other vegetables are intensively grown in plots averaging less than one acre, separated by banks or hedges of gorse, tamarisk and escallonia with hardly any woodland, heath, marsh or normal large fields of arable and pasture. In these conditions in early September, I found a virtual absence of the flocks of field birds such as rooks, larks, buntings, lapwings, gulls, and woodpigeons which dominate farmland in coastal areas of southern England and much of northern France. Even jackdaws and starlings occurred only in smallish parties of under a couple of dozen. Dunnocks and wrens proved to be the commonest small birds, followed by blue tits and robins, migrating chiffchaffs and meadow-pipits, linnets and blackbirds. This unusual selection gives some indication of the species which are most favoured by an exceptional development of garden conditions on the ground without much tall cover. Gardens as we know them in England are inhabited by some birds which are interested in the cultivated floor but not the cover, others which use the cover but not the cultivated floor, and some which need both.

This brief review must serve to indicate the complexity and variety of the group of garden and garden-like habitats, and the pace at which fresh types are evolving. What all these types have in common is that they are wholly or largely planted and more or less continuously and intensively tended by people, while most of them are so much used that birds living in them must be capable of tolerating fairly close human approach without feeling bound to fly away. A frequent, although less general feature is that the birds of this group of habitats are under an increasing amount of human notice, mainly friendly, and that in many cases efforts are made to provided them with food or food plants, water, nesting-boxes and a certain degree of protection, so that they become a wild population partly 'farmed' by man, in much the same way as partridges or duck, but without the same ulterior motive of sport.

One effect of this attitude is that garden birds tend to become less wary and more approachable, in so far as they can do so without being promptly selected out by cats, or by small boys with catapults. Another is that gardens create conditions exceptionally favourable to high population density and thus tend to modify or in extreme cases to break down territorial habits appropriate in less sheltered conditions. For similar reasons garden birds tend to sing earlier and longer than individuals of the same species in other habitats—a point which is very noticeable with such a loud singer as the throstle. Also, the margin of advantage which garden birds enjoy is greater in winter, especially in a severe winter, than in summer, creating a tendency for birds belonging to other habitats to shift into gardens and similar places in winter, or at least in hard weather. For these and other reasons the study of birds in gardens as compared with other habitats is a promising and important field for research.

Most garden species are resident in Britain and are to be found in

PLATE XIX.

Eric Hosking

Pair of pied flycatchers

Eric Hosking

(*a*) Greenfinch

G. K. Yeates

(*b*) Pair of cirl-buntings

gardens at all seasons, but the spotted flycatcher, whitethroat, lesser whitethroat, blackcap, willow-warbler, wryneck, turtle-dove and a few other summer migrants are fond of large and sheltered gardens or garden-like places. No winter migrant, except the rare and irregular waxwing, shows a similar preference.

While the species most characteristic of gardens are described individually in a later chapter it will be convenient to review briefly in the following pages the birds most closely linked with gardens, orchards, hedgerows and kindred habitats, and to notice the nature of their dependence.

The crows as a family are dependent on this group of habitats in Britain only to a certain extent for nesting-places and in very limited conditions for food. Carrion-crows are fond of nesting in commanding and sometimes quite low hedgerow trees, and they also nest and roost in city parks, although in these cases much of the food is sought elsewhere, such as on the mud of tidal rivers. I have however seen one in Battersea Park pick up and fly off with half a roll of bread, which it dropped near me. Rooks nest freely in trees which happen to stand in the gardens of suburban or country houses, but they have no other claim to be regarded as garden birds. Jackdaws nest in a wider variety of holes in trees and buildings, again frequently in gardens, and they will at times forage in chicken-runs and on vegetable plots, but this is quite incidental and comparatively insignificant among their sources of food. Magpies are particularly fond of breeding, and also of birds-nesting for eggs to eat, in wild hedges; on the Continent they often nest in gardens but if they ever did so in Britain the habit has been largely eradicated by persecution. Jays, although wary woodland birds by origin, are now following the example of the crow rather than that of the magpie, by colonising large suburban gardens and even the inner London parks, where since 1940 they may often be seen tamely feeding or drinking, and even rooting about for food in litter baskets.

Starlings nest freely in suitable holes in buildings or trees in gardens and will also come in foraging in some numbers, but the vast majority feed in the fields.

Greenfinches are very typical birds of our larger gardens, especially where there are thick shrubberies and ornamental trees, and also of such garden-like habitats as churchyards, cemeteries, and shrubberies on bypasses or reservoir banks.

Goldfinches settle in only the more spacious and mature gardens or blocks of gardens, and in some public parks and many orchards. Lesser redpolls are also garden birds locally in very favoured places, as in parts of Surrey where large gardens have been carved out of the birch heaths on which they live, and in Connemara where I have found them in the fuchsia hedges around the houses. Linnets again overflow freely into gardens of houses adjoining commons, or on cliff-tops and other places where they flourish. The serin, one of the most successful garden birds on the Continent, has not so far spread to Britain, although many breed within a hundred miles of our shores. Bullfinches come into gardens and orchards frequently, and sometimes breed in them, but are preponderantly still woodland birds, and are not often found far from some sizeable wood or thicket. Crossbills have diminished a great deal as breeding species in England during the past twenty years, and have disappeared from a number of localities; their liking for old conifers has often led to their foraging and even breeding in gardens in suitable places.

Chaffinches, although not fully treated in this book because they are predominantly woodland birds, are very fond of gardens wherever they contain enough mature trees. Even the central London parks attract a few chaffinches, and no species is more capable of becoming tame enough to take food at close quarters, even hopping onto tables, where people eat out of doors, to take crumbs off a plate in front of them. Like starlings, our native chaffinches are normally resident, but are reinforced in winter by a big immigration from across the North Sea.

Hawfinches are mainly woodland birds, but will breed in large gardens in wooded areas, and even in public parks; they are notoriously fond of raiding the green peas.

Of the buntings the corn-bunting is to some extent a hedgerow bird, but only in so far as convenient perches happen to occur along hedge-tops rather than on isolated bushes, fences or telegraphs. The yellowhammer is one of the birds most attached to hedgerows at all times of the year, but is hardly a typical garden or orchard species (although it may overlap into these from neighbouring habitats) and shows a marked dislike for getting cut off from open country. The cirl-bunting is the only member of its family in Britain which definitely likes gardens, and is often to be found nesting as well as feeding and singing in them. It seems curious that the scarcest and most retiring of the English breeding buntings should be so much a garden bird.

All the same, few would detect its presence, in the ample gardens which it best likes, but for the persistent and almost unmistakeable song.

House-sparrows are found in the hedgerows chiefly where they adjoin farms or buildings, and especially after the breeding season; they also swarm over many gardens and garden-like habitats. Tree-sparrows are more garden birds on the Continent than with us, but they do come freely into gardens and even breed in them, especially where marshes with pollard willows or other specially suitable sites adjoin. Larks are only exceptionally garden birds, although I have known a woodlark which sang from the wire-netting round a Surrey tennis-court and another which lived in a neglected public garden on the edge of open country. Much the same applies to pipits, the tree-pipit being liable to occur in gardens adjoining suitable commons or open woodland, and the rock-pipit occasionally in gardens on suitable coasts, while the meadow-pipit frequents market-gardens on passage.

The pied wagtail is attached rather to houses and farmyards than to gardens, but is a familiar sight on the lawn wherever nesting conditions are attractive, and will also visit pools and streams in gardens freely. Some fortunate gardens with fast-flowing streams are able to attract the graceful and lovely grey wagtail, which I have even seen swooping to explore the small goldfish pool in the closed-in courtyard garden of the Naval and Military Club in Piccadilly.

While tree-creepers will visit trees whether they happen to be in gardens or not, and I have found one's nest against a Surrey garden gate, they remain woodland birds neither attracted nor repelled by people and their interference. The introduction of Wellingtonias from California into the British Isles has, however, proved enormously popular with tree-creepers looking for roosting places, and may even help to reduce winter mortality by diminishing deaths from exposure on cold nights. As these trees are usually planted in parks and pleasure gardens rather than in woods they may also exert some attraction to tree-creepers to extend their beats in these directions.

Nuthatches are definitely fond of large gardens, parkland and old orchards, where there are well-grown broadleaved trees such as oaks, beeches, limes, walnuts and sweet chestnuts. They are southern birds, mostly found between the Channel and the Trent and Mersey; their population has a curious outlier north and south of the Tees. They are readily attracted to nesting-boxes, and will come to bird-tables for nuts, when they not infrequently get reported as kingfishers by unwary bird-lovers. The nuthatch is a cheerful active bird with a striking variety of loud whistles, pipes, trills and hollow metallic utterances, all very loud and clear, so that without being in the least aggressive he naturally comes to dominate his favourite gardens and to make himself the spirit of the place.

The tits, all woodland species, have found little trouble in taking our gardens into their ambit, being untroubled by any serious shyness and being fond of the mixture of trees and shrubs of different sizes and varieties which we choose to grow. The willow and marsh-tits are the most conservative, only normally coming to gardens which adjoin their woodland or thicket habitats, and which contain similar types of trees, although they are regular in plenty of gardens where these conditions are fulfilled. Long-tailed tits also come into gardens rather sparingly where these link up with their normal haunts, and they have not become dependent on us either for winter food or for nest-sites, their only frequent act of exploiting our presence in Britain being when they use our poultry as a convenient source for the vast number of feathers with which their nests are lined.

Coal-tits, having a strong preference for conifers, are attracted in southern Britain to the neighbourhood of gardens, parks and similar places where so many of the more mature conifers have been planted during the past two centuries. However, with the spread of afforestation with coniferous trees, the proportion of the English coal-tit population in gardens is likely to be reduced as the strength of the species grows. Any conifer from anywhere suits a coal-tit; at Hillsborough, County Down, I was glad to find Irish coal-tits flitting about very appropriately in the avenue of Irish yews.

Great and blue tits are outstandingly the most garden-loving of the family, and are common in almost all the group of habitats discussed in this chapter throughout the British Isles except in parts of the extreme north. They are in fact among the most characteristic of all garden birds, but being even more typical of woodlands do not fall to be treated in detail in the present volume. Great and blue tits are both highly adaptable species which have learnt to take advantage of us in many ways. Our various patterns of artificial woodlands, including gardens, orchards and hedgerows suit them at least as well as the natural types, and some of the trees and shrubs which we plant

PLATE 21

Robert Atkinson

Blue tit returning to nesting hole

PLATE 22

Eric Hosking

Goldcrest feeding young

are especially fruitful sources of insect or vegetable food. In addition they are extremely fond of bacon-rinds, coconuts, fat and other food which until the wartime fat shortage used to be put out on bird-tables or hung dangling for them to eat in full view of our windows. The cutting off since 1940 of this substantial source of high-grade winter food must have been a serious deprivation to both species, and it will be interesting to see whether, when the supply can be resumed, it will lead to any decline in the practice which has spread during the same period among these birds of attacking the lids of milk bottles and helping themselves to some of the cream. The first record of this practice was from near Southampton in 1921, but it has only become widespread and common since 1939, and may have been stimulated by wartime conditions which cut standards of living and stimulated the search for substitute foods among tits as well as among people. (See Pl. III, p. 32).

More recently an even more curious and inconvenient habit of entering houses and tearing paper has spread like wildfire through many parts of the country. Notices have been ripped down, covers torn off magazines, not even sparing *British Birds*, and even lampshades and cartons on dressing-tables have been hacked about. The British Trust for Ornithology, which had already assisted a study of the interference with milk-bottles, has appealed for records of this surprising new habit to be notified to Colonel W. M. Logan-Home of Edrom House, Edrom, Berwickshire, who is conducting an investigation into it. So far as is yet known it breaks out in early autumn and persists until early spring, and the fact that it has occurred where the birds are regularly fed at bird-tables suggests that hunger is not responsible, although the possibility that a deficiency of some special sort of food, such as fat, may be behind it cannot yet be ruled out.

In addition to their other forms of dependence on people blue and great tits are particularly apt to use artificial nesting holes, and particularly wooden nesting-boxes in which they breed freely. Pied flycatchers also take readily to nesting-boxes, and are comparable with cirl-buntings in their readiness to settle in suitable gardens in the relatively restricted regions of Britain where they breed. There have been welcome indications lately that their breeding range is expanding southwards and eastwards from their strongholds in Wales and the North of England and the Scottish Lowlands. They are not however such pronounced garden-lovers as spotted flycatchers, which are familiar in gardens, parks, cemeteries, recreation grounds and similar places in summer in most parts of the British Isles.

Goldcrests, like coal-tits, are drawn into gardens and similar habitats not by any liking for human society but by their fondness for many of the evergreens which we plant, not even barring monkey-puzzlers. No tree is too high for them and they will also come down at times even into the undergrowth, but tall stands of mixed evergreens are their most favoured haunts, even where the number of trees is small. They rarely show any sign of noticing our existence, yet contrive on the whole to keep well out of our way, and have no direct dependence on us either for food or for nesting-sites. The firecrest, a rare autumn and winter visitor, shows rather more preference for gardens, and less for conifers. Most warblers may be seen in a garden at one time or another, the garden warbler being one of the most unusual in this habitat, as one might expect from its name. The willow warbler, blackcap, whitethroat and lesser whitethroat are the only species which show any marked and widespread tendency to take up territory in gardens, and then only in those which are unusually large and sheltered, or closely adjoin more suitable types of country. We have no warbler which is so typically a garden bird as the icterine and melodious warblers are over much of the Continent. The disappearance of the human race from these islands would perhaps most inconvenience the lesser whitethroat, which is fond of tall dense hedges, especially those containing occasional tall ash-trees, and the whitethroat which likes lower and more straggling ones, but the warblers as a family have probably less of an interest in our survival than any other large passerine group, even the currants and raspberries which we grow being no more than a local and seasonal luxury for a small minority of them.

On the other hand the fortunes of the thrush family are very substantially bound up with the continuance of gardening and the planting and tending of amenity trees, shrubs, hedges, lawns, flower-beds and vegetable plots. The throstle, blackbird and robin are outstanding as garden birds, and as such are fully discussed on later pages, together with the mistle-thrush which although much less numerous is also a characteristic inhabitant of larger gardens and of garden-like habitats. Two members of related families, the dunnock and the wren, are equally familiar and widespread in our gardens. All of these fend very largely for themselves, and usually choose more or less natural nest-sites, but their numbers and densities have been increased and their British distribution modified by the spade-work which gardeners have done for them. All these birds can be said to follow the gardener's spade, led by the robin which hastens to sit on it the moment the gardener's hand is removed. The nightingale is often heard close by gardens and would, like the blackcap, probably settle in them more freely if gardeners could exercise more self-restraint in cleaning up untidy undergrowth.

Woodpeckers are only incidentally garden birds, although green woodpeckers ornament many lawns by hopping up and down them in full view and barred woodpeckers are almost as liable to be seen on trees in orchards, parks and gardens as in woods. Even pied woodpeckers not infrequently visit large gardens, sometimes breeding in them.

Wrynecks are included among the species fully treated in the following chapter (pages 164–68) on the strength of the fact that when they were well-known British birds they especially haunted gardens and orchards, although the number still to be found in these or any other situations is becoming pitiably small.

Of the owls the little owl is fond of orchards and parklands rather than normal gardens, and the tawny owl is primarily a woodland bird which does not hesitate to roost, breed and forage in gardens where they sufficiently resemble or adjoin woods, and which does not disdain to nest in owl-boxes put up for the purpose, but is otherwise in no way dependent upon people.

No day-hunting bird of prey is in any sense a garden bird, although both kestrels and sparrowhawks will use hedges and parks for hunting. Although herons will sometimes come and clean up a goldfish pool one early morning and mallards have been known to rear young in window-boxes and tubs of irises, neither the herons nor the ducks and geese can be considered as garden birds, nor in fact can any other group on the British list except the doves. Of these the woodpigeon, although it breeds in woodlands and forages in the fields has become a breeding species in some wild undisturbed country gardens and in hedgerows, cemeteries and churchyards. It has also, losing its country shyness, colonised a number of town parks and gardens, especially in London, where it feeds as well as breeds. The stock-dove, a bird which nests in holes and is therefore not driven towards thick woodlands for breeding cover, is notably fond of parkland and of hedgerow and shade trees with suitable holes, but has so far colonised town open spaces only in small numbers. The turtle-dove likes to breed in thickets rather than in tall trees, and therefore takes fairly readily to large gardens, but has not yet become sufficiently trustful to stand close approach or frequent disturbance.

As this outline shows, the birds of gardens, orchards, hedgerows and like habitats, are predominantly passerine species, and the development of these habitats, which is rapidly proceeding, can hardly fail to strengthen still further the already immensely strong position of the leaders in this group.

M. D. England

Spotted flycatcher and supporter

PLATE XXII.

Throstle and red squirrel at feeding table

Dr. M. S. Wood

CHAPTER 6

LIFE-HISTORIES OF GARDEN AND HEDGEROW BIRDS

A DISCUSSION of the nature of gardens, orchards and hedgerows and of their bird life in general was attempted in the previous chapter. We now come to consider more fully the life-histories of twelve species specially characteristic of this group of environments in Britain, namely the :—

GREENFINCH	*Chloris chloris* (L.)
GOLDFINCH	*Carduelis carduelis* (L.)
YELLOWHAMMER OF YELLOW BUNTING	*Emberiza citrinella* (L.)
CIRL-BUNTING	*Emberiza cirlus* L.
SPOTTED FLYCATCHER	*Muscicapa striata* (Pall.)
MISTLE-THRUSH	*Turdus viscivorus* L.
THROSTLE OR SONG-THRUSH	*Turdus ericetorum* Turton
BLACKBIRD	*Turdus merula* L.
ROBIN	*Erithacus rubecula* (L.)
DUNNOCK	*Prunella modularis* (L.)
WREN	*Troglodytes troglodytes* (L.)
WRYNECK	*Jynx torquilla* L.

Two or three other species characteristic of gardens but given to nesting on buildings are considered at length in Chapter 8.

GREENFINCH
(Pl. xxa, p. 109)

Greenfinches are well suited to life in and around low thickets. Their dull greenish plumage is highly protective, their thick strong bills can grind to pulp seeds and kernels, and they find nesting-sites as well as food, roosts, and shelter among the branches and foliage. They are accordingly among the most specialised of our smaller birds. They cannot breed, and can hardly comfortably exist for long without ready access to vegetation, which must grow over a yard high, yet on the other hand must not consist wholly or mainly of tall trees. They are very fond of evergreens, especially yews, cypresses, ivy and young conifers of Christmas-tree size, but they also like hawthorns, elders, hornbeams and many kinds of ornamental and orchard trees. In tall or extensive woodlands they keep mainly to the fringes and clearings. A botanic garden or arboretum, the gardens and grounds of a large house, churchyards and cemeteries and orchards, thickets and tall hedgerows, amenity belts and plantations are among their favourite strongholds, and where one greenfinch is found there will usually be several or plenty, even in the breeding season, when they nest in loose colonial settlements. They are lowland birds, my highest record being at about 780 ft. on the South Downs.

Despite these specialised and restricted ways of life, greenfinches are neither skulking nor stay-at-home, and they often fly far afield at a considerable height. When I was operating the Oxford Ornithological Society's trapping station in the middle of Christ Church Meadow I was surprised by the strong attraction the greenfinches found in the pinhead oatmeal with which the trap was baited. Although the site was some way from cover some greenfinches visited it so often as to become a nuisance, and they would return, sometimes within a day or two, even if taken off and released between six and nine miles away in various directions. One was retrapped twenty-nine times in seven weeks, and another thirteen times in one week. These and many other trapping records show that the greenfinch is more inquisitive than wary, and finds it difficult to resist favourite foods. Although foraging widely over a district our native greenfinches appear to be largely resident in the same places throughout the year. They are greatly reinforced during the winter by immigrants from the Continent, where greenfinches of one race or another stretch right across to China and Japan, and from the Arctic forests down to the Mediterranean and North-West Africa. In the British Isles they are widespread in suitable country.

Greenfinches feed much on the ground, under trees or bushes, in roadways and in stubble fields. Occasionally they may be seen feeding on ripe grain, clinging precariously to the tops of the slender stems. The young are fed partly on insects and spiders, but the diet of the adults is almost entirely vegetarian, consisting largely of seeds and berries.

The greenfinch's gait on the ground is hopping, and the wing-beats in flight are interrupted at intervals, giving an up-and-down motion, yet strong and quite fast. The flight-note is softer and gruffer than the chaffinch's, and intermediate between it and a linnet's, but definitely distinguishable from both by a trained ear. More striking notes are the very clear musical "tu-we-ee", often uttered from dense cover with mystifying effects, and the drawling unmusical "zeeaire" commonly heard between March and August. The song is uttered from high in a tree or often on the wing in a curious slow-motion circling song-flight, with fanned tail and deliberate wing-beats. It is a rather formless jumble of repetitions of the normal notes grouped into brief phrases and uttered in a pleasantly leisurely manner, without assertion or challenge.

In Germany I have found the nest on the pollarded stump of a willow and up the bole of a poplar, as well as in ivy on a building, and I have seen one built almost entirely of wool and another with a piece of newspaper built into it. In England, however, the nest is usually in thick bushes or shrubs, and is built of twigs and moss, lined with roots, hair or feathers. Four to six whitish eggs with fairly light reddish-brown and violet markings are laid in late April or May; they hatch in about a fortnight and the young, which are fed by regurgitation, fledge about a fortnight later. There are two or three broods.

The upper-parts are greenish, russet-olive or grey-brown, the wings dark brown or blackish with a broad lemon yellow bar; there is also a conspicuous lemon panel each side of the tail; the old cock has a lovely yellowish-green breast, but on hens and juveniles it is brownish-grey or greyish; the bill is pinkish-white and the feet fleshy. The length is just under 6 inches, and the build stout.

GOLDFINCH

(Pl. 23, p. 128)

Goldfinches are among the most attractive of our birds, bright in colour, elegant and lively in action, and musical in voice. They further recommend themselves by their fondness for feeding on the seeds of the thistle, groundsel, knapweed and other plants objectionable to both the farmer and the gardener, and to some extent on insects. They are hardy enough to survive even our most severe winters without disaster; in Germany I have seen one in fine condition and plumage after three days of snow lying with eighteen degrees Fahrenheit of frost. They are trustful, but just wary enough to take care of themselves, and they readily adapt themselves to changed conditions.

With all these merits and assets it might be expected that goldfinches would flourish exceedingly in England but, although far from scarce, they remain thin on the ground in many districts. In the nineteenth century their charm led to their undoing, since as popular cagebirds they were netted wholesale by the many birdcatchers who were then active. Wild Birds Protection Acts, which had little effect on gamekeepers, hit the birdcatcher hard and gradually the trade in wild-caught cagebirds was almost crushed out, to the benefit of the goldfinch perhaps more than of any other species. Yet it is doubtful whether this control of birdcatching has been anything like so important a benefit to the goldfinch as the decay of British agriculture since 1870 and the march of thistles, ragwort and other weeds over what was once well-farmed land.

Recently we have seen great and successful efforts to restore British agriculture, but up to the present these efforts have not led to a conspicuous decline in tall weeds. The pressure to plough and crop unsuitable land has often helped weeds to spread, particularly where the effort has had to be abandoned, and even on good land clean crops are far from universal. On grassland no serious attempt has yet been made to reclaim large acreages of poor or rough pasture, and in the many areas infested by rabbits thistle and ragwort have it all their own way. Much land also has been taken for defence, for airfields, for communications and other uses which often seem to involve laying down a good part of the site to thistles.

So far, then, the policy of reviving British agriculture has been no less favourable to goldfinches than the policy of letting it rot. The longer prospect looks more uncertain for goldfinches, and in some respects much less good. Greater use of science, improved equipment and education and higher prices make for more intensive farming and may be expected to involve more wholesale eradication of weeds, and the elimination of waste patches, banks and even hedges on many farms. The ploughing and reseeding of grassland aims at results which will leave little food for birds such as goldfinches. On the other hand the quicker pace of development and more frequent transfer of land from one use to another creates a large shifting margin of excellent foraging areas, and as goldfinches prefer gardens, orchards and parkland for breeding their position must be in some ways strengthened by the spread of these habitats. While goldfinch numbers and distribution are therefore likely to be unstable and changeable in many districts, and should be closely watched, there is at present no reason for alarm.

One of the best illustrations of the adaptability of goldfinches and their responsiveness to change is their recent history in Inner London. A century ago the builder and the birdcatcher between them made the London area unhealthy for goldfinches. There is a record of twelve dozen of them having been caught in the nets in a single morning on the site now occupied by Paddington Station. About seventy to eighty years ago Harting and other good naturalists described them as rarely seen and not nesting in Middlesex even beyond the suburbs. Early this century they apparently nested near London only to the north, and up to 1940 they visited Inner London only very occasionally in winter, usually singly. During the war they began to be seen fairly freely even in the Green Park and Hyde Park on weedy patches above bomb shelters and among the anti-aircraft guns. In October 1940 a flock of about forty flew over St. John's Wood, and a year later several were seen in Battersea Park. In 1942 I saw one over Chelsea and three young were seen at Earl's Court in August, and in May 1943 one was seen over Charterhouse Square. In 1945 I saw one in Queen Mary's Garden, Regent's Park in March and Sir Philip Manson-Bahr saw two with three young at the same place on 17 June, which may well have been bred in the park.[1] In August I saw two feeding on thistles in Green Park within fifty yards of Piccadilly and the Ritz Hotel, and in September there was a party of a dozen near the bandstand in Hyde Park; others were seen quite frequently in Battersea Park or over Chelsea. With the clearance of the war sites visits became less frequent after 1945, but in April 1949 I found a nest with one bird sitting and two others near it in a crowded part of Battersea Park—apparently the first recorded in Inner London for a century. The bird sat closely for over a week, but on returning from an absence abroad I found it had been taken. Goldfinches however continued to haunt the park and succeeded at a later attempt in rearing young. Thus the recovery

[1] He informs me that he heard young goldfinches twittering by the open-air theatre there during a performance in July 1949 and considers that they now breed there regularly, although he has not actually found a nest.

of a songster which half-a-century ago seemed lost to London for ever now looks within grasp.

Goldfinches are sociable, unaggressive birds and band together outside the breeding season in flocks of several dozen, often mingling with linnets, greenfinches, siskins, and other relatives. Even in the breeding season they manage to disperse without asperities and to visit neighbouring pairs freely.

The distribution of goldfinches is in patches rather than a continuous block; the patches are largest and thickest round much of the coast and in suitable garden, orchard, and upland areas, becoming fewer in Scotland, where goldfinches are entirely absent from most of the outer islands, and in Ireland, although goldfinches are in good numbers in some parts right across to the Atlantic coast. Most of our birds move about for relatively short distances, and they do not share the linnet's habit of leaving for France in large numbers in the autumn, nor do many immigrants come here. On the Continent the related subspecies penetrates north only as far as the Trondhjem area of Norway and the Leningrad area of Russia, whence goldfinches of one form or another cover virtually all the rest of Europe and overlap into the nearest parts of Asia and Africa. In a transect in the French Alps I found goldfinches from the valley floor at 2,370 ft. up to an altitude of about 5,300 ft., stopping short at the same height as the chaffinch and linnet, and rather lower than the yellowhammer. In Britain goldfinches are not much met with above the 1,000 ft. contour.

Like tits goldfinches gather their food in all sorts of postures, even hanging upside down from thistleheads or fluttering with the beautiful broad gold banners showing across the open black wings. Also like tits, and unlike most finches, they are not so fond of feeding on the ground. They are at home in trees both tall and small, especially in summer, and are fond of cemeteries and churchyards; at Selborne I have seen one perch on the headstone of Gilbert White's grave. On the ground they hop; on the wing they have a charmingly impulsive dancing motion, circling round singly or in parties or flocks, and undertaking long cross-country excursions unusally freely for so small a bird, with every appearance of enjoyment.

Although not remarkably powerful or varied the song of the goldfinch is unfailingly pleasant to listen to. The liquid easy notes pour out in gay and melodious streams of sound, gently rising and falling, more expressive of the joy of living than of challenge to rivals. The song is often uttered on the wing and seems perfectly to convey, like the flight itself, the gay impulsiveness and contentment of the singer. When delivered in chorus, as it often is, it puts into music the natural and intense delight of goldfinches in one another's company. It is a song entirely free from harshness, from aggressiveness and from imitation. It is at its best from March until June. The call-notes are like separate drops of the stream which composes the song, and have the same delicious liquid quality, yet they carry warning of the bird's approach over longer distances than many louder and more incisive cries.

Goldfinches' nests are even neater than chaffinches', showing the same effective use of moss and lichens, and of vegetable down and wool. They may be fairly low in bushes or high in trees, and either in a fork or well out in the smaller twigs. The five or six eggs are bluish-white with few but bold reddish-brown streaks and some ashy markings; they are laid mainly in late May, and take twelve or thirteen days to hatch, after which the young take slightly longer before they fly. There are two, or occasionally three, broods. As the young are fed by regurgitation, by both parents, they do not need to be fed so often and in such small beakfuls as most small birds.

Both male and female share the same bright plumage, with the odd vertical broad stripes of crimson, white and black on the head and face. The black wings and tail have each feather tipped with white, and the striking golden-yellow broad band makes a vivid wingpanel at rest and extends the whole length of the wings in flight. With all these strong contrasts the eye almost overlooks how much of the plumage is buff or brown or dull white. The stout pointed bill is large for the bird and very pale. The young are streaky, with linnet-like heads, but have the typical wings and tails. Goldfinches are plump birds, but very small, being under 5 inches long, and weighing only half-an-ounce, or about 14 grammes—roughly the weight of one wren and one goldcrest together.

YELLOWHAMMER OR YELLOW BUNTING

(Pl. 24, p. 129)

Buntings are apt to be patchy as well as widespread in their distribution, and the yellowhammer, although the commonest of the group in Britain, is no exception. In areas where yellowhammers are well established they easily come to be regarded as prominent members of the bird population of any country district, but those who go about with their eyes open find themselves constantly crossing invisible lines, on one side of which are plenty of yellowhammers and on the other few or none. These lines are less distinct than those bounding the tribes of corn-buntings, for the cock yellowhammer although as conspicuous in plumage is less dominating in voice and in choice of singing posts, and is moreover sufficiently widespread and common to be easily taken for granted unless the observer takes trouble to notice where yellowhammers are not as well as where they are.

Superficially the yellowhammer appears simply to require a fair amount of grass or arable land adequately enclosed with hedges, banks or woodland edges, or alternatively open rough land with a fair amount of scrub. By origin it seems to be a scrub bird which has probably increased by taking advantage of the artificial scrub habitat provided by hedged fields. As these hedges are a comparatively modern development the probability is that much of the increase took place as recently as the seventeenth century. But there is plenty of land of this sort which looks right and yet is avoided by yellowhammers. In one case I noticed that they were common on one side of a main road along the edge of the chalk hills, but were hardly ever seen the other side, on the valley levels. In another case, in Cornwall, the coast is occupied by corn-buntings on the more exposed stretch and cirl-buntings on the more sheltered, while yellowhammers only become frequent a mile or so inland. On Lough Corrib in the west of Ireland I found yellowhammers common round the shore while on the islands suitable-looking grassy habitats with scrub were occupied instead by reed-buntings. In the mountains of Kerry I noticed that along the roadsides yellowhammers ascended well beyond the limit of cultivation in the passes.

Although yellowhammers are fond of uplands I have never found them at above 1,000 feet in England or Scotland. In a transect of the French Alps however, the yellowhammer proved to be the only bird present all the way up from the valley floor at 2,370 ft. to above the tree-limit at about 5,500 ft. They are not normally found in old woodlands, but cling on in tall thicket or in forestry plantations long after the growth has become uncomfortably dense for them. There is some shift in quarters from grassland to stubbles and rickyards in winter, and on an Oxfordshire farm where a number of observers

C. W. Teager

Throstle sunning itself

John Markham

Blackbird's nest with eight eggs in peasticks, Norfolk

co-operated in carrying out censuses over several years, yellowhammers proved to be fairly stable in summer population, but numbers in winter averaged nearly double the summer level, and were occasionally much higher. In the breeding season yellowhammers were outnumbered more than four to one by blackbirds, more than two to one by chaffinches and throstles, and also slightly by linnets and robins among the residents and by whitethroats among summer visitors. Such a result is probably fairly typical, yellowhammers being often in the second half-dozen species in order of local abundance, but not often in the first. Although they are gregarious outside the breeding season their flocks are usually of moderate size, gatherings of much over a hundred being unusual in most parts. They mingle freely with other buntings and with finches.

The only parts of the British Isles in which yellowhammers do not occur regularly are the Outer Hebrides, although they are contracting their breeding range in Orkney and are only passage birds in Shetland. Few if any species occur in such numbers both on the extreme east coast of Suffolk and also the extreme west of Co. Kerry and Connemara, and from the Isle of Wight to Sutherland. Continental birds can be seen passing down the east coast in autumn, and are thought to winter here. In Europe yellowhammers breed from the Arctic Ocean to Spain, Portugal, Italy and Yugoslavia, and in Asia a different subspecies extends from Siberia to Iraq and Persia, while in winter migrants penetrate as far south as North Africa.

Yellowhammers feed mainly on the ground, hopping and pecking as they go, at corn, seeds and leaves. They are among the very few creatures other than man which are known to eat blackberries at all freely. Insects, earthworms, slugs and spiders are also taken. They spend much time in the hedges, and not infrequently perch on trees and overhead wires; more rarely on buildings. The flight is lighter and more graceful than a corn-bunting's, with often interrupted wingbeats and undulating motion; long and fairly high flights are commoner than with most small birds. They roost in loose order, often in the thick of a hedge; when I flushed three in a Buckinghamshire lane by moonlight some hours after dusk in January, I was surprised how they rocketed vertically upwards for about four feet above the perch, before flying off with the usual alarm.

In the same place in February I watched an extraordinarily furious duel between two cocks in a field, with a third bird, apparently a hen standing by, very close and excited. They sprang up a few inches from the ground, scrapping with wings, beaks, and apparently feet as well, the blows being many and loud. After two brisk rounds they flew to a coppice hedge for a third, one in the heat of the fray uttering the normally drawling song in quick time but without the least abridgement. In the end one of the cocks allowed himself to be driven off, leaving the other in full possession. Although looking rather phlegmatic, as buntings do, they are excitable and somewhat quarrelsome birds. Once in Germany, when I caught a fledgeling which a cock was feeding on the ground he became extremely excited and gave a good performance of "injury-feigning".

The nest is usually on or near the ground under herbage, shrubs or young trees, often in a hedge or bank, but sometimes in a bush. It is quite neatly built of grass and moss lined with horsehair; the earliest I have found was in Germany on 23 April and the latest in Ireland on 10 August; even September and October nests have been recorded, two to three broods being reared. The clutch is small, usually three or four eggs, whitish with often a tinge of purple and marked with the most extraordinary fine scrawls in dark brown, which used to earn it the country name of "scribbling lark."

The loud "twick" flight-note is one of the most familiar country bird calls, and the song is also familiar partly owing to the unforgettable but very imperfect rendering "A - little - bit - of - bread - and - no - cheese". It is monotonous in its regularity (typically starting about every tenth second and lasting about three, of which one is taken up by the "cheese") in its tone (although considerable individual variation can be detected) and in pattern, the main variation being the fairly frequent dropping of the "cheese" ending, when caution becomes necessary to avoid confusion with the cirl-bunting's song. The voice is of unusually high frequency. Full song starts in February and continues till August; it is rare in autumn.

The sub-song, which I heard from an apparent juvenile in Brittany in August, began with the "twick" call and continued with scratchy twanging notes arranged like a faint and fumbling imitation of chaffinch song, without enough accent on the conclusion. It was in pattern more advanced than the full song, but carried only about 15 yards.

Yellowhammers are a size larger than house-sparrows except in the bill, being about 6½ inches long, but weigh rather less, averaging under one ounce; they are fairly stoutly built forward, but taper towards the tail. Individual cocks vary a good deal in the extent and brightness of the yellow on the head and breast; the hens are duller and more smudgily marked, and the young show little yellow, but all adults have very warm reddish-chestnut rumps and white outer tail-feathers, which are the best identification mark and often the most easily seen as birds fly off in front. The upper-parts are chestnut-brown with many broad black streaks, and there is often some reddish on the chest.

CIRL-BUNTING

(Pl. xxb, p. 109)

A map of the Roman Empire would give a tolerable idea of the distribution of the cirl-bunting, which is predominantly a bird of the Mediterranean, but spreads some way across Asia Minor, up the Danube Valley and through the mildest parts of western Europe. Like the Roman colonists it conveys an impression of tolerating best the most benign and genial spots in this dank northern island, in which it reaches both the most northerly and the most westerly points of its breeding range. Yet it is of all our buntings the least given to migration, being nearly stationary at all seasons, and its power of surviving severe winters compares favourably with that of most of our passerine birds.

In England cirl-buntings breed in some strength only within a comparatively narrow belt of roughly sixty miles in depth behind the English Channel coast, and in the lower Severn basin and near the coast of North Wales. Very small numbers may be found in a few other South Midland and Welsh counties, but north of the Trent and Mersey, in Scotland and Ireland, and even in East Anglia, cirl-buntings are less often seen than several accidental wanderers from the Continent. Nesting was, however, proved in Lancashire in 1941, and in Leicestershire also during the war, and there have been several recent occurrences in the north of England.

Even within their chosen part of Great Britain cirl-buntings are only to be found in certain districts, usually within a very few miles of the sea or of some specially favoured stream or hill escarpment, and within these districts again there are special spots for which they show continuing preference. I have known such spots in Sussex, in Surrey, in Hampshire, in both North and South Devon, in Cornwall and in Gloucestershire, and I have also watched cirl-buntings in the breeding season in Middlesex and Herefordshire as well as abroad, yet it remains extremely difficult to define their exacting tastes. Their choice seems to fall not on a particular type of country but on the kind of spot to which a discriminating man might wish to retire after spending a good deal of his life in some much warmer climate than ours. In fact near Worthing, by Spithead, in the South Hams of Devon, in the Cheltenham-Malvern area and elsewhere, cirl-buntings select the same neighbourhoods and even frequent the actual gardens which such people have picked. Often, although not always, the chosen spot is on a slope, sometimes facing south, and in any case not too exposed; unless the slope is gentle and does not rise too high, the cirl-bunting is likely to be found fairly near the foot of it. In Devon I have seen a fine cock bathing for several minutes in a small clear stream, splashing the water over himself and preening, and nine years later another was seen bathing with enjoyment in an ornamental garden pool not far away. The frequent choice of a territory near to water may be no accident.

The cirl is the most arboreal of our buntings, and is probably never found living far from some well-grown trees. A preference for elms has often been suggested, and elms are certainly often used as singing stands, but that may follow naturally from their attachment to the less exposed places and their frequency as shade and hedgerow trees commanding meadows and lanes. I have not myself been convinced of any special preference, although I have several times observed the cirl-bunting closely associated with Monterey cypress, better known as *Cupressus macrocarpa*, and other exotic trees and shrubs which flourish in some of its haunts. Once in Selborne churchyard I saw one settle on the tip of a cypress tree. One singing in Devon in mid-May was fond of new-leafing oaks, and sometimes took time off to walk about the branches like a chaffinch, eating defoliating caterpillars.

Cirl-buntings need a blend of well-grown timber, undergrowth or hedges, tall herbage and some open ground, and are therefore attracted to the fringe between different types of country, such as the foot of the hill, the last garden next to the open fields, or the lane between the wood and the farm. With such requirements they could hardly become common birds, but at least they have the luck to depend on a habitat which is being deliberately cultivated and spread by civilisation, rather than one which is vanishing. It might, therefore, be expected that the species would gradually increase, and this may well be happening. The cirl-bunting has, for example, been established in Selborne

Eric Hosking

Goldfinch's nest in cypress

Eric Hosking

Cock yellow hammer feeding young under dead bracken

village throughout the past hundred years, but it was unknown to Gilbert White, who could hardly have overlooked it had it been there in his time. So unobtrusive and retiring is it, however, and known to so few reliable observers, that we have insufficient evidence to guide us on the trend of cirl-bunting numbers.

While cirl-buntings often occupy adjoining territories they are usually well spaced out, and a cock will shift about between singing stands a fair distance apart. Nevertheless the daily orbit appears unusually restricted considering the size and sturdiness of the bird, and this is borne out by the length of the song-period which is exceeded by few British species, and those mainly the most sedentary. It almost appears as if territories are rejected unless they can satisfy the bird's varied needs for most if not all of the year. Territories surveyed by R. P. Gait in south-west England in 1946 measured about nine or ten acres each, and appeared to be separated by strips of no man's land.

The cock being so conspicuous both in dress and in voice, and the territory so often close to cultivation or even dwellings it might be expected that the cirl-bunting would be hard to miss, but the opposite is true. An observer unable to recognise or unlucky about hearing the song, might live for longish periods in the midst of cirl country without being aware of the fact. Even a cock singing continuously may take some minutes to find, owing to his frequent and infuriating habit of sitting for long periods on a perch buried in the midst of a tree's leafy canopy, where he is difficult to make out from above, below or sideways. This habit is all the more inexplicable since on other occasions the singer will sit boldly on the topmost twig of a tree or bush, or sometimes on telegraphs without appearing to mind the exposure.

As buntings go the cirl is a pretty unsociable bird, although it is not uncommonly seen in mixed winter flocks or parties with chaffinches, or mingling momentarily with linnets and other species. The largest gathering I have found was in South Devon, where I watched a dozen or more, the sexes about evenly divided, feeding on ploughland in mid-October, close to where Montagu first discovered the species in Britain in 1800.

Although by no means silent or skulking, and making frequent sorties for food from the cover which they love, cirl-buntings when not singing usually manage to be less obtrusive than almost any other bird of their size. On a perch the usual posture is slanting and neither graceful nor alert, but the slenderer build and longish, narrow notched tail save them from looking as lumpish as some other buntings. The flight also is rather light for a bunting, but hesitant and unimpressive. The behaviour generally often has a lethargic and unenterprising air about it.

The cirl-bunting is one of the few British species for which occasional song by the female has been proved. It sings regularly through half the year, and more or less frequently in some part of all the remaining months, from September to February, and has a remarkable and baffling range of song variations.

In form the song is almost always a simple monotonous quick repetition of about half-a-dozen to a dozen brief notes, all at the same pitch and volume, giving a phrase of mechanical rhythm like a fast-moving light shuttle, and of only about 2 seconds' duration. A cirl timed over 10 minutes in Surrey delivered 29 songs, with a rate of four a minute, apart from interruptions. Another timed in Devon in May, however, achieved 6 songs a minute three times, 5 twice and 4 four times, giving an average of nearly 5, and a cock in North Cornwall in August kept up a rate of 6 or even 7 songs a minute for several and often many minutes on end.

Despite the simplicity and uniformity of its basic pattern close study shows that the actual content and substance of the song, even as delivered by a single individual cock, may be extremely variable. Sometimes, perhaps most often, the phrase consists of clear but flat notes, stridulous and devoid of ringing or warbling qualities, but pleasantly free from grating or shrill tones. Sometimes a harder variant is used in which the notes are not unlike rapid percussions on stone. Another variant consists of somewhat scratchy, scraping, shrill notes and it is this which comes close to a yellowhammer's song without the prolonged *eese* or "cheese" note at the finish. Yet other variants use whirring and clicking notes which sound each much faster and more complex than those already so imperfectly described, and which in quality come close to a louder, briefer snatch of a grasshopper-warbler, or sometimes to the whirring alarm of an excited wren. All these variants—and there are doubtless others—conform to the general pattern in volume, duration, and evenness of emphasis throughout, but careful comparison shows them to be very distinct in content. An individual singer will repeat one variant for a time and then switch to another, and I have known a single cock try at least five in turn over a period. Some sing most repeatedly in the morning and evening, but there is often a good output in the middle of the day, although there may be long intervals of silence. In the more mechanical and vibrant variants the song is accompanied by a rapid up-and-down shivering of the tail, the body also sometimes quivering in sympathy as the singer sits otherwise motionless, without shifting the perch for minutes on end.

The call-note is an urgent, anxious rather low but penetrating and husky *zeet* or *zeet-zeet*, sharpened on alarm into a brisker *zit-zit*, while the young have as a hunger call an almost grasshopper-like and very penetrating *zett-zett*, running together in the excitement of being fed into a very rapid *zett-zett-zett-zett-zett*.

The untidy large nest is built near or on the ground of coarse grass, moss, and sometimes leaves (in France I have found wool in one), and is lined with horsehair and fine grasses. The cirl is a late breeder, rarely laying before mid-May, and rears two or sometimes three broods, sometimes having eggs as late as September. They are marked with almost black streaks and hairlines on a bluish or greenish ground, and usually number 3 or 4 in a clutch, being incubated by the hen in about eleven to thirteen days, after which the young fledge in a similar period. The cock is said to feed the young only occasionally but I have seen one feeding them in the nest in France, and feeding fledgelings in Cornwall.

In size and build the cirl-bunting is slightly smaller and slenderer than the yellowhammer, and is in the main a dull sparrow-like bird on which is superimposed, in the case of the cock, a brightly coloured and exotic head and breast plumage. Between the black chin and blackish olive-green crown the cock shows yellow cheeks crossed by a conspicuous blackish stripe; the lower throat and neck are also yellow, and a broad delicate olive-green band runs across the breast and backwards up on to the nape: below this follow large rounded reddish patches and yellow underparts. All these the hen and young lack, being brownish above and yellowish streaked with brown below, but all share the characteristic greenish-brown rump colour, in contrast to the chestnut-rumped yellowhammer. Moulting cirl-buntings in August and September are a sad sight.

SPOTTED FLYCATCHER
(Pl. XXI, 25, pps. 116, 132)

Living almost entirely on winged insects captured in flight the spotted flycatcher ranks with the swallow and martins, the swift, and the nightjar, but its methods are totally different. Instead of cruising about to seek the flies, midges, gnats, butterflies, moths or other insects on which it preys, the flycatcher picks some commanding perch and waits for its food to come along. Its design for living is only less economical than the spider's; its plumage, its posture, its haunts and its migrations are all determined by that design.

Mouse-brown above and dull white below, with darker markings on the folded wings and with striations to break up the outline of the head and breast, the spotted flycatcher is almost strikingly inconspicuous. Only the longish straight broad-based bill, the unusual and businesslike bristles surrounding it, and the bright black eye disclose what underlies that innocent appearance, while the feeble, tucked-away legs proclaim a bird given to moving only by air. Longer-winged, shorter-legged and less plump than a robin, the spotted flycatcher is basically similar in size, but individuals vary and a large pied flycatcher may even be bigger than a small spotted.

The spotted flycatcher's posture is equally revealing—alert, concentrated, thrust forward slantwise with the tail continuing the line of the back, the breast puffed out stoutly and the head hunched firmly down on the shoulders, all ready for the sudden out-and-home forays by which it collects its food.

Small flying insects are most abundant and regular near good shelter and cover, and it is at such points that the flycatcher takes up territory, especially in the grounds of country or suburban houses, in well-planted churchyards, in parks or open woodland and along well-planted roadsides. Although often nesting in creeper or in hollows on houses, or in other continually disturbed situations, it rarely becomes tame enough to ignore human comings and goings, or to visit its nest without regard to the risk of giving the site away. Its close association with people is evidently an unwelcome necessity to it.

Willingly or not, no bird has become more woven into the stuff of English summer life than the spotted flycatcher. It is he who sits hunched on the wall of the cricket-ground under the elms and horse-chestnuts as the match goes on, or who guards the boathouse above the waiting punts and canoes, or who flits across the lawn, as we have tea out of doors, to a nest in the espalier on the old wall, or who darts from the old grey tombstone as we pass through the lych-gate into the country churchyard. But for an austere refusal to court attention and

Eric Hoskins

Spotted flycatcher entering old kettle

Eric Hosking

Young mistle-thrushes

popularity nothing could prevent him from being one of our best-known and best-liked birds.

Unlike the very localised pied flycatcher the spotted flycatcher penetrates to nearly every part of the British Isles, and is to be found throughout the summer in most districts except the extreme north of Scotland, Orkney, Shetland and the outer Hebrides, where in normal years it is seen, if at all, only on passage. Yet comparing lists made in various parts of the country it is my impression that the main weight of spotted flycatcher population is in the midlands, south, and west, and that it has a distinct tendency to be more local and to occur in smaller numbers in the east. Undoubtedly among the European and Asiatic flycatchers the spotted is the most westerly in distribution, breeding from the borders of India and from Turkestan to the Atlantic, as far north as the shores of the Arctic Ocean and southwards into North Africa, and wintering in tropical and southern Africa. Although they have been known to appear in early April or even March, spotted flycatchers are not often seen with us before May, and it may be after the middle of that month when the main numbers arrive, with stragglers and passage birds still coming in until June. Although some begin to leave in July they are much in evidence throughout August and will sometimes stay until after the first autumn frosts, the latest date on which I have seen them in England being 23 September, when I saw one in Selborne churchyard and another in Wolmer Forest. It has been proved that the same individual bird, as in other migratory species, will return to the same spot in successive years.

Spotted flycatchers are exceptional in being strongly territorial small birds; of inconspicuous plumage, which have entirely failed to develop a territorial song. Like hawks and shrikes they spend almost their whole waking life in a kind of patrolling activity which enables them incidentally to chase off any intruders and no doubt renders specialised territorial warnings and challenges superfluous for them. The fact that their favoured habitats are so limited and so readily kept under observation must also help in simplifying the problem of ensuring proper dispersal. They are also watchful and jealous about other intruders. I have seen one swoop at and drive off a young robin from tiles just above its nest, snapping its bill in the pursuit as loudly as if catching flies, but being quite content as soon as the robin had retired from the immediate neighbourhood of the nest.

As summer advances and insects become more plentiful spotted flycatchers grow almost sociable, banding together in family parties and even in larger loose companies, especially as migratory movement develops, and early in September I have seen at least sixty hawking in Kensington Gardens within an area of not much more than an acre. In this area there was a very populous wasps' nest but even when they perched within a few yards of it the flycatchers took no notice of the wasps. They will however take bees on occasion. The backs of park chairs, like posts or fences or walls, make satisfactory look-out posts, as well as buildings and the barer outside branches or twigs of trees. They are fond of low perches, sometimes only a foot or so above the ground, and will quite often settle on the grass to pick up an insect, although the great bulk of their food is taken in quick swerving dashes and hovers on the long pointed wings. In August I have seen a young bird descend to the ground and begin picking up straws and stems of dry grass, carrying them about for a minute or two and dropping them.

There is little display, but the male will feed the female in spring, and I have seen a presumed male excitedly moving from branch to branch singing the confused jumble of excited shrill jerky notes which passes for a song with this species and finally addressing himself in a crouching attitude to a presumed female; this was early on a June morning in Maidenhead Thicket. The song, which is of about the same order of development and of melody as a house-sparrow's, is a rather wheezing rambling medley reminding one of the excited cries of birds engaged in some kind of mild scuffle, and is so weak as to be inaudible at quite a short distance. It usually embodies the characteristic "tzee-tucktuck" anxiety note. This is often described as the alarm, but it is uttered on the least provocation when the bird seems to suffer merely mild anxiety. When really excited and angry it will use instead a harsh churring almost shrike-like "tchaar" accompanied by frequent spasmodic half-opening of the wings. Young birds make a constant shrill affirmation of their perpetual hungriness, using a version of the normal shrill high-pitched "tzee" call-note.

The rather neat small nest of moss wool and hair is often bound with cobwebs, and may be anything from a yard to a good many feet up, usually against a wall or treetrunk, or both, but often also on a beam or in some old nest; I have found one in an elder fork, in Germany. There are usually four or five handsome pale greenish eggs with plenty of rather robin-like reddish markings. They are laid in the second half of May or early June, taking 12–14 days to hatch, and the young fledge in a similar period. Sometimes there are two broods, and even three have been recorded; eggs in July or early August are not exceptional.

MISTLE-THRUSH

(Pl. 27, p. 140)

The mistle-thrush belongs to the fringes between wooded and open country. By artificially multiplying such fringes and by preventing the vegetation from developing to its normal climax man has powerfully aided the mistle-thrush, which must have been a scarcer bird before parks and gardens and orchards and hedgerow trees were spread across the face of the land. With the growth of suburbs and the planting of berry-bearing and fruit trees for ornament or use the outlook for the mistle-thrush is increasingly favourable. It might almost seem that the unconscious object of the most enlightened landscape gardeners and town and country planners is to create as many mistle-thrush territories as possible.

Owing to this fortunate preference for some of the types of habitat which happen to be most fostered and protected by man the mistle-thrush is now widely distributed in the British Isles. The spread has been most conspicuous in Ireland, where the species was not known until the beginning of last century, but is now freely seen right across from the Irish Sea to the Atlantic promontories of Connemara, and may be found breeding even on small islets in the lakes of the west. This century it has also colonised the Outer Hebrides at Stornoway. Although typically a lowland bird in Britain the mistle-thrush penetrates well up into the hills in suitable places; on the Shropshire border in early June I flushed more than fifty at an altitude of nearly 1,500 ft., but these no doubt had moved up after breeding nearer sea-level. At this stage in the year family parties join together into flocks of several dozen birds and shift to hills, downs, moors and even mountains where they enjoy the summer plenty of insects, and become for the time largely ground-birds, until the autumn frosts destroy their seasonal food-supply and drive them back to the lowlands. There is some emigration to France and Ireland and some immigration and passage movement from north Europe, in addition to weather-movements in winter, but in many places the birds stay all through the year. In France I have seen mistle-thrushes, on suitable territories just like those they would choose here, right down to the Mediterranean, but towards the Alps they tend to occupy a zone intermediate between high and low altitudes; for example in a sectional census in Dauphiné covering from 2,370 to 10,500 ft. above sea-level, all ten of the mistle-thrushes encountered were around 4,000 ft. up, at the highest point where walnut trees grew.

In appearance and in habit the mistle-thrush is one of the hardiest and toughest birds we have, not flinching from laying eggs in early spring weather which resembles winter at its worst, and singing with undaunted if not actually enhanced high spirits in the most forbidding conditions of snow, rain, high winds and even darkness. During the eclipse of the sun on 24 January, 1925 when most birds were perceptibly cowed by the sudden darkening, I noticed in Buckinghamshire that the mistle-thrushes continued singing unrestrainedly. Yet mortality is evidently high, judging by the fact that two broods have to be reared in a season and that the annual contribution of young birds for replacement of losses is not so much less than in a frail species such as the long-tailed tit. This was borne out by the fact that after the very severe weather of early 1947 mistle-thrushes virtually disappeared from a number of areas including some as far west as Wales and Cornwall, while some apparently more delicate birds suffered less.

Mistle-thrushes have few obvious enemies, and are well able to look after themselves. No bird preserves territory more jealously, or is more vigilant and stout-hearted in driving out predators known or suspected, including hawks, owls, crows, magpies, jays and cats. The grating "rough music" with which the mistle-thrush accompanies its assaults on such intruders is one of the most familiar bird sounds of spring in mistle-thrush country. The boldest mistle-thrush I ever knew built a large and conspicuous nest, visible fully 150 yards away, in the Green Park in 1947. It was in a fork less than 6 ft. up in a little tree within a yard of a busy path and within 100 yards of Buckingham Palace and one of the busiest traffic centres in London. There were four young, but it was impossible to count them unless the bird was away as it merely crouched on approach and when I actually touched it, pecked my hand. On one occasion I found near the nest an astonished man with a largish brown and white dog on a lead. As he moved near the nest one of the mistle-thrushes, entirely ignoring his presence, swooped repeatedly down at the dog from a height of some 20 ft. to within a foot of the dog's nose, the dog (still held on the lead) snapping back without effect. This happened about ten times in quick succession.

I spoke to the man, who was under the impression that the bird had taken some unaccountable dislike to his dog and was quite unaware, like almost all who passed that way, of the nest just in front of him. The pair were fearless in taking food to the nest even when people were near. They foraged quite close, largely at any rate on the open grass. Unfortunately, shortly before they should have fledged, the young were found lying dead under the nest at the end of April, probably through human action.

Even in London, where it flourishes, the proportion of mistle-thrush to blackbird population rarely reaches one-seventh, and in many districts it is much lower, even before taking into account the mistle-thrush's much more restricted habitat. It is difficult to attribute this either to failure to compete effectively, or to enemies or other forms of mortality, and the implication is that the bird's requirements are so exacting and the territory which it demands so extensive as to impose a fairly strict limit on its numbers. Outside the breeding season the mistle-thrush is more sociable than the blackbird or the throstle, but much less so than the fieldfare or redwing. Although loose parties or small flocks of mistle-thrushes are not rare, especially in summer, they behave very much as individuals, often going off in different directions when disturbed, and they seem only accidentally to mingle with other species, being less inclined that way than almost any other field birds, except game.

The mistle-thrush is by no means an adaptable bird and in all essentials its way of life probably remains unchanged since prehistoric times. Its dependence on civilisation, although close, is almost wholly indirect. Bold and fearless as it is, the mistle-thrush is among the least tameable of birds, and does not choose to live in any closer sight or sound of man than its natural way of life requires in these man-infested islands. Even in London, where it is one of the few species which has continued since time immemorial to breed in the most central Parks, it remains a wholly wild bird, disdaining all artificial aids to its diet and nesting, and not allowing itself to be closely approached on the ground. An interesting possible clue to the mistle-thrush's tastes is that in London it distinctly prefers the waterless Green Park, which is the least attractive to most species. The Green Park is more sloping than the others and contains groves of low hawthorn tree and one large clump of mountain ash, which is the favourite centre for the mistle-thrushes. Is it possible that the pleasure and good living associated with these berry-bearing trees influences the mistle-thrush to haunt their neighbourhood and even to breed in them at seasons when they have nothing special to offer?

In addition to parks and gardens in town and country, and orchards which are among its favourite haunts, the mistle-thrush is also at home in woodland clearings (or even in woods, where the high canopy is not too dense) and in coppices, spinneys, hedgerows with trees, and wherever on heath or hillsides a remnant or likeness of the primitive "savannah" habitat of trees dotted about open country is still to be found. Whether the trees are native or exotic, coniferous or broad-leaved makes little difference so long as they are self-sown or planted in the style of the gardener rather than in the style of the Forestry Commission. Water appears to have little attraction for the mistle-thrush, but is by no means avoided.

On the ground the mistle-thrush stands high with a stiff, proud, alert carriage, remaining immobile for considerable periods, which are followed by determined rushes usually all in the same general direction. Unlike our other native thrushes it never seems to mind how far it is from cover, and is hardly ever found in the undergrowth or ditches so much favoured by them. Even perching on low hedges is unusual, and it is fondest of the crowns or upper branches and trunks of trees rather than the thicker foliage. Artificial perches, particularly buildings are generally avoided, although posts and wires for transmitting electricity are used to some extent.

The flight of the mistle-thrush is high, wide and handsome. It habitually rises to a hundred or more feet up, and excursions of a quarter to half a mile or more are everyday events at most seasons. A series of deep strong wing-strokes on a rising curve are followed by a moment's pause during which the bird dips almost as deeply as a green woodpecker, although the flight gives a freer and more bounding impression. In spring a pair will engage in chases from tree to tree; in Hampshire in March I once saw pairing take place almost immediately the hen alighted, the cock fluttering onto her back and treading her twice in quick succession, with fast-beating wings. Under excitement also the song is uttered in flight.

All our thrushes combine the ability to deliver a highly musical song and a highly unmusical alarm note, but none more markedly than the mistle-thrush. Whereas the blackbird's alarm usually covers a hurried retreat into dense bushes the mistle-thrush's harsh chiding churr more often serves as a war-cry, summoning all like-minded creatures to drive out some common enemy. This cry is also used in a less strident form as a flight-note when disturbed. The song is loud, wild and clear, flung out with a defiant ring and uttered with just a trace of the roughness of voice which comes out in the alarm. It is a short phrase, normally of about five or six notes, not unlike the blackbird's in general pattern but distinguished by its bawling instead of leisurely delivery and its rough instead of mellow tone. The mistle-thrush is the largest of our birds to develop what is ordinarily called a song, and its performance is easily the loudest warbling utterance to be heard here. It reaches full volume not long after the New Year, and falls off badly in May, although I have heard it in every month except July and August and there is sometimes a full-throated resumption in early winter; for example, one was singing particularly well in St. James's Park on 20 November 1947, as the crowds hurried past for Princess Elizabeth's wedding. The only subsong I have heard was like the true song but imperfect and very much fainter.

The mistle-thrush has the rare virtue among mainly fruit-eating species of being fond of almost all those fruits which we dislike, and of hardly any which we cultivate. The seeds or berries of rowan, yew, hawthorn, holly, mistletoe, juniper, rose, ivy and bird-cherry, and almost all the other common wild fruits except the blackberry, are taken with immense zest as the birds cling to the slender twigs with spread wings, often fluttering madly to keep a footing, and so far forgetting their wildness as to let themselves be approached within four or five yards without much difficulty. The mistle-thrush's inadaptability and its habit of moving to the hills in numbers just as the main fruit season comes on, and returning just when it is nearly over no doubt help, but whatever the reason it is a remarkable piece of good fortune that this fine and capacious bird should levy so light a toll on our fruit crop. Slugs, worms, spiders and many insects are also eaten, and killing of young birds is on record.

The nest is built in March, or a little earlier or later, usually in a tree, and with a preference for the fork, although sites out on lateral branches are not uncommon, especially in yews. It is a big, solid, typically thrush-like structure of moss, bents, roots and grasses with a mud core embedded under a thick lining. The eggs, normally four, are almost as large as a jay's and usually handsomely marked with rich brown blotches and spots. Incubation is completed inside a fortnight and fledging hardly less quickly, so that young can be seen flying about in numbers during the last half of April. Two broods are usual.

In length of bill, legs and tail the mistle-thrush is similar to a blackbird or fieldfare, but it is a markedly larger and longer-winged bird, although no heavier than a large blackbird. The upper-parts are generally grey-brown, the underparts creamy much spangled with bold rich brown spots, some much darker than the rest and some in patches. The underwing shows white in flight, and so do the tips of the outer tail-feathers. The rump is sometimes a brilliant yellowish-brown. The young birds have a very white-headed and generally mottled appearance, showing much more white than the adults and having a quite different stupid and cloddish bearing, which is apt to prove confusing to those unfamiliar with them. (See Pl. 26, p. 133)

THROSTLE OR SONG-THRUSH

(Pl. XXII, XXIII, 28, pps. 117, 124, 141)

The throstle, as I shall persist in calling the song-thrush, is in normal periods almost certainly among our ten most numerous birds, yet is outnumbered by the blackbird with remarkable consistency in almost every habitat which it occupies and in almost every part of the country where it occurs. Both species are found throughout the British Isles, and both range through almost every type of woodland and have spread wherever there are hedges, gardens, orchards or almost any sort of leafy cover with access to open feeding grounds. In this competition to expand over as much ground as possible the blackbird is definitely the more enterprising, pushing higher up the hills wherever there is any cover to be found, and colonising remote islands and other unpromising places where the throstle does not penetrate. In the Hebrides, however, which have their own peculiar subspecies (*Turdus ericetorum hebridensis* Clarke) the throstle really becomes a heath-thrush and almost vindicates the unfortunate Turton. It was he who a hundred and fifty years ago wrongly supposed that he had found a new species of heath-thrush which he christened *Turdus ericetorum*, and thus became the unwitting instrument of misnaming the throstle four generations later, after its previously accepted scientific names had been overturned on more or less pedantic grounds. These Hebridean birds not only live out on the moors and peat-hags but are also fond of the seashore.

PLATE 27

Eric Hosking

Juvenile mistle-thrush and hawfinch at drinking hole

PLATE 28

T. M. Fowler

Throstle sitting

In places on the mainland in summer throstles are found above the thousand-feet contour and beyond the tree-limit, but they are predominantly lowland birds, flourishing best in comparatively sheltered and even lush conditions, and taking years to recover after a severe winter. It may be said as a rough generalisation that the richer and more varied the habitat the more throstles it is likely to hold and the more narrowly will blackbirds outnumber them. On the less rich habitats blackbirds will often come out at from two to five to each throstle, and on really poor ground the throstle is apt to drop out entirely. This even holds good of difficult habitats in the middle of towns; in the inner London parks blackbirds outnumber throstles by up to 8, 10 or even 20 to one and in some close residential areas blackbirds are well established while throstles are missing. In winter, when the climate makes every British habitat so much poorer, many parts of Scotland and northern England are entirely deserted by their throstles, and this applies not only to the hilly and exposed districts but to mild and sheltered areas, for instance on the coast of Ayrshire. A census carried out twenty years ago by my brother on a small wooded estate near Ayr showed no throstles in January, 8 in February, 9 in March and 36 in April, while there were 30 blackbirds in January and only 35 in April. On a sheltered Surrey estate I found a similar situation; at the end of December there was only one throstle on forty acres against 21 blackbirds, although in the previous breeding season there had been 14 adult throstles against 25-26 blackbirds. The report by Lack and Venables on the habitat Distribution of British Woodland Birds, indicated blackbird majorities of from two to four to one in various kinds of woodland. On the other hand on an Oxfordshire farm with good hedgerows and feeding areas the Oxford Bird Census found a fairly stable winter population of throstles usually not much lower than the breeding population.

Ringing shows that throstles from Great Britain winter in the Isle of Man and Ireland, and on the Continent from Belgium and Luxemburg through France to Spain and Portugal, while birds of the continental race make various movements through the British Isles and birds hatched in Holland and Belgium have been found in England and Wales in winter. To add to the confusion, birds of the so-called British race have been found nesting on the nearest parts of the Continent from Holland to West France and at Stornoway in the outer Hebrides, while birds breeding in Shetland have been claimed to belong to the Continental race. The Hebridean race, unlike both the British and Continental, appears not to be migratory. The Continental race breeds almost throughout Europe except on the shores of the Arctic Ocean and the Mediterranean, and extends well into Siberia, birds from the north wintering down to the borders of Arabia, the Sudan and North Africa.

In many parts of the Continent the throstle is by no means the common and universal species that it is with us, and although found in towns in France and Germany it is on the whole much more a woodland form.

While blackbirds may often be seen bunched together in some numbers in winter at favourable points throstles appear distinctly less sociable, although in very favourable areas in the breeding season pairs may be pressed fairly close. The very loud far-carrying territorial song is more fully developed than in any other thrush known to me, and lasts for a longer period of the year, being at fairly full strength from early December until July, and heard with varying frequency in the remaining months, most rarely in August. The song has a special character and structure of its own, being neither a continuous warble nor a simple pattern of notes, but a complex prolonged composition in which each note is usually repeated with the briefest of pauses some two, three or four times. The range of frequency in the throstle's voice is wide, from as low as a cuckoo to nearly as high as the top notes of a wren, and it has a shrill clarity and persistence which dominates even a loud background noise of traffic, or the more pleasant distraction of a wood full of other singing birds in spring. I have found one singing vigorously from the ground up against a haystack, but generally the singer chooses a stand well up in a tall tree, from which the song can carry to its full range of a quarter of a mile or more. The call-note is a rather tense "sipp", usually uttered as the bird flies up to a safe perch, while when alarmed it uses a voluble harsh chatter, more scolding and less ringing than a blackbird's. The throstle is a good mimic and I have heard it imitate a partridge so well as to get a reply from one several hundred yards away; I have also heard imitations of other species, including redshank and nightingale.

Throstles feeding on the ground stand stiffly upright, head cocked to one side, alert for signs of prey near the surface, running or hopping at intervals for a few feet to the next listening point.

Throstles are fond of cover, particularly in low trees, bushes and hedges; although strong and direct fliers they are not often seen making excursions of any great length. They feed largely on snails, which they smash in quantities on stones or other convenient hard objects, and on earthworms which they drag out with great energy and spirit. They also eat spiders, centipedes, beetles, flies, moths, earwigs, ants and seasonally also many berries and fruits. Some of these, such as holly, yew, elder and hawthorn, are of no interest economically, while others such as cherries, raspberries, strawberries and red currants, bring the throstle and the fruit-grower into sharp conflict. The loser in this conflict is usually the fruitgrower, who can only console himself that throstles, unlike starlings, do not gather on his fruit in large bands from all sides, and that such good as they do in eating pests is also done in their home orchard, and not in some distant field belonging to someone else.

The British Trust for Ornithology Nest Record Cards for the throstle have recently been analysed by Miss E. T. Silva, and they give a much fuller picture than was previously available of the breeding habits of the species. Throstles in Great Britain usually lay clutches of four or five eggs, between mid-March and mid-July. The clutches increase from an average of 4 eggs at the end of March to nearly 4½ in late April and early May, falling back to 4 in June and under 3½ in July. Clutches in Holland average almost one-third of an egg more, while in Finland they average half-an-egg more. Unfortunately it is not yet clear whether the clutch-size varies in different parts of Britain, but records of hundreds of broods over ten years, indicate a tendency for at any rate June broods of young to be larger in north-west England than for the country as a whole. The size of broods also varies considerably from year to year, a good year for young throstles being usually a good year for young blackbirds as well. Of the total eggs laid it was found that 71 per cent hatched and 55 per cent gave rise to fledged young. Eggs are laid daily till the clutch is complete, but hatching is spread over a period. In one nest which I watched closely in Sussex in 1917 the first chick hatched on 7 April although the fourth and last egg had only been laid on 26, or at any rate after the late afternoon of 25 March, a period of only about twelve days. The fourth chick was not hatched until two days after the first. The only one which survived left the nest on the afternoon of the 19 April, only ten days after the last egg had hatched, but twelve after the hatching of the first. The average period for hatching is about 13½ days and for fledging just over 13 days. Two or three broods are reared, yet even this high rate of reproduction is not enough to enable the throstle to keep up its numbers in face of the mortality which it meets in severe winters.

The very solid nest of grass and other material with its strange hard basin-like mud lining is usually in a bush or low tree, but at Selborne I found two on the ground in the steep banks, and I have been shown one in which there were two broods running in a fruit tree trained against a house just under a bedroom window. The beautiful blue eggs with their few black spots are well known to almost all who have ever looked at birds' nests.

Throstles are about 9 inches long, a good deal smaller and generally lighter than blackbirds, but their legs are longer proportionately. The upperparts are more or less warm brown, with a slight reddish tinge; the breast and flanks fairly warm buff, paling to white on the belly and brightening into golden-brown under the wings. Except on the lower throat and upper breast the dark, bold, spade-shaped or arrow-headed breast spots are erratically placed and widely scattered.

BLACKBIRD

(Pl. 29, p. 144)

During the breeding season the blackbird is almost certainly more plentiful than any other bird in England except the chaffinch, and with its conspicuous and familiar habits, large size, loud voice, memorable plumage and almost universal distribution, it is possibly the best-known of all our birds. Best-known, that is, if seeing and recognizing amounts to knowing, for when we check the extent of our scientific information about the blackbird it is still astonishingly patchy and limited.

From the middle of Westminster to the Atlantic coast of Connemara and the moorlands of Hoy, I have rarely managed to find any considerable area of the British Isles where the blackbird was not one of the common birds, ranging from sea-level up to 1,350 ft. on the Welsh border and to fully 1,400 ft. in the Highlands, and penetrating into nearly every habitat from the seashore in the Scillies to small islands in lakes and wherever trees, shrubs or even tall heather can give cover. Perhaps only windiness deters the blackbird more than its competitors, and in some exposed coastal districts in Devon, the Scillies, and the

Cock blackbird and young

Robert Atkinson

John Markham

Robin

Outer Hebrides it takes second place to the throstle, while in St. Kilda it has no foothold at all. It breeds right up to Unst in Shetland, and on the Continent a little farther north in Norway, but avoids the colder northern and eastern parts of Russia, although various races extend across to China and southwards to Kashmir, Persia, Syria and North Africa. The blackbirds which I have seen in other parts of the range such as the Azores, the Crimea, France, Germany and Austria have usually seemed less enterprising in their choice of habitat and more inclined to restrict themselves to the most favourable and sheltered spots than our blackbirds in the British Isles. They have, however, spread out from the woodlands and colonised towns and gardens in the same way as here. Our breeding blackbirds are much less prone to emigrate for the winter than our throstles, but some certainly shift to Ireland and France while many immigrants come in from Scandinavia, the Low Countries and Germany, of which some stay while others continue their journey across the Channel.

While robins are the most numerous British birds which are wholly unsociable, blackbirds seem to be forced by sheer pressure of numbers to tolerate blackbird company rather than to welcome or recognize it. It is especially when large supplies of berries are available at one spot that blackbirds can be seen gathered in dozens, but they still behave as a chance assembly of individuals and show no sign of developing social action, except possibly when they mingle with more social species such as mistle-thrushes or fieldfares, and when they gather for display as described below.

Blackbirds are plumper and more rounded in shape and less stiff and tense in carriage than our other thrushes, with a high-mettled nervousness easily touched off by the slightest incident, but quickly releasing itself in a torrent of chatter or in a dash for cover, which every blackbird likes to have close at hand. The rather loose-cut wings and the handsome spreading tail are freely used to perform graceful dancing movements as the blackbird advances, pauses and listens, or alights on a perch or on the ground. Sometimes, in a stiff breeze, the tail is cocked straight up like a great wren's; sometimes, in preening, it is depressed at right angles and fanned out like a kestrel's for the orange bill to attend to. At the end of preening I have seen a cock repeatedly wag his tail with surprising force and rapidity from side to side, never up and down, at the same time shivering the wings about a quarter opened, like a cuckoo. In courtship particularly the cock fans out his great handsome tail and bows his head, dipping the orange-tawny bill. Although normally a walker he is capable of moving in swift bounding hops.

It looks as if with the blackbirds song is less important than dance, and the fact that the song begins so remarkably late in February and is so little heard in the preceding six months, has been plausibly attributed to the part played in winter by the communal gatherings of blackbirds. Miss Averil Morley has described how these gatherings occur at fixed points from November until April, during most of which time only the females keep territory, and attempt to drive out birds of both sexes. Gatherings are attended usually by a handful of birds—I have seen as many as twenty—and in some cases, especially in London and other towns the birds assemble from quite long distances. At first these gatherings are most notable on going to roost, when their clamour impresses the least observant. Often they are at a drinking and bathing place, or in the garden of a city square; I have known exceptionally a gathering place in tall trees, but normally open ground near cover seems to be essential. The main activity at these gathering grounds is a very primitive dance, consisting of formal mock combats and chasing, or persistent shadowing of another bird, with bowing demonstrations and a good deal of clamour and chatter, the males being usually in a strong majority. About Christmas, and two clear months before the song becomes frequent, cocks can be observed beginning to court the hens, or in some cases hens courting cocks. According to Miss Morley the hens have considerable trouble in ridding their territories of rival hens, and it is only at the end of February or the beginning of March that the cocks begin seriously to drive out others. This coincides with the general resumption of song, about three months after the throstle, two months after the mistle-thrush and over six months after the intensely territorial robin.

Perhaps because it is held back until the fulfilment of mating and territorial possession, instead of being a challenge and a signal from the start of these ordeals, the song has a mellowness and contentment above all others. Each fluting phrase lasts only about two to four seconds, and is normally uttered from six to nine times a minute, with pauses of about the same length between each utterance, which may be heard in good conditions as far as half-a-mile away. Once on a sunny March afternoon with thick snow underfoot I heard an extraordinary variation from a cock about 70 ft. up in the crown of a tall plane in Battersea Park. It was more throstle-like than blackbird-like, including nothing recognizable as typical blackbird song, and quite lacking in blackbird finish and mellowness of execution. The most characteristic and frequent phrases were a very clear wild triple "Yehudi, yehudi, yehudi" with the clarity of a greenshank's call, followed by the alarm given in a slurred and unaccented form, so that when first heard it sounded like some other species imitating a blackbird; there were also some throstle-like repetitions of single notes up to five times, and a very throstle-like double "tee kur, tee-kur, tee-kur" followed by a more chuckling "gurdy, gurdy, gurdy". The whole song was appreciably less loud than normal, although carrying some way, and was obviously only half-formed, giving an interesting demonstration of common basic material out of which the very distinct songs of the blackbird, throstle and mistle-thrush may well have evolved. Fortunately for the inexperienced listener, such confusing medleys are of the utmost rarity.

The alarm of the blackbird is much more than a note, having the length and pattern of a song, but not the function or the music. Beginning with two or three throaty chuckling protests it suddenly rises in pitch and accelerates into a shrill excited chatter, to which as an anti-climax two or three more chuckles are often added. The chuckle is also used by itself as a note of caution, and the familiar "mik-mik-mik" expresses suspicion and emotional tension on such occasions as going to roost in winter or mobbing an owl or cat, which is a favourite sport of blackbirds. Another less common note is a shrill thin "seee" with a family likeness to the redwing's. Quite distinct from this, and used by both sexes in spring is a somewhat drawn-out anxious "eeeee" of low volume but of exceptionally high frequency. It is a pure and unaccented piece of sound delivered up to 12 or more times a minute often in a still trance-like attitude, sometimes after or during a skirmish.

The song is usually given from well up in a tree, or sometimes from housetops and while flying; more rarely from the ground.

Unlike our other large thrushes blackbirds have no great fondness for high or long flights, but they are quite powerful on the wing and remarkably adroit in plunging at full speed through small holes in fences or hedges, or rocketing up vertically to clear the face of a quarry or similar obstacle. Flight-speeds so far recorded range between about 26 and 33 m.p.h.

At most seasons blackbirds feed chiefly on the ground, often in the open not too far from cover, but also under trees and bushes, where they sometimes make as much noise as a man walking, by turning over the dead leaves with their bills and feet. In the second half of the year wild and cultivated fruits and berries attract them greatly, and the young birds fly in good time to join their parents in gorging themselves on strawberries, raspberries, currants, cherries and other fruit, passing on to mulberries, pears, apples and figs in their season. When these are finished elder and ivy berries and many others are eaten, as well as insects and worms.

Blackbirds make large rather shallow nests in which the core of mud is concealed by a thick outer cup of grasses and moss and an inner grass lining. Normally it is built in bushes or hedges a few feet from the ground, but in cat-infested towns, especially London, it is commonly much higher in a tree, creeper, or even on the crook of a drainpipe under the eaves of a house, while other sites are favoured locally. Near Selborne many nests are on the ground, and I have found four in this position within a hundred yards, and on another occasion half-a-dozen on stumps in the sides of the sunken lanes. I have also noticed this habit at Highcliffe-on-Sea on the Hampshire coast, but in many areas ground nests are uncommon. In Germany I have found one on the ground under a pile of brushwood, and nests in stacked brushwood are not infrequent. (see Pl. XXIV, p. 125). The bluish green eggs with their rich reddish-brown and grey markings are laid from March or April onwards, there being usually from three to five in a clutch. Two to three broods are reared, even four having been recorded. The hatching of the eggs and fledging of the young each take about thirteen days.

I once saw a hen blackbird in Surrey at dusk fluttering as if crippled along the path in front of me, falling over and tumbling repeatedly in complete silence; another time when I took shelter from a hailstorm a blackbird went off her nest by me and made off with great clumsy strides for about ten yards before taking wing—a clumsier version of "injury-feigning".

David Lack has analysed the mortality of blackbirds from ringing returns and has found that with a possible life-span of twenty years the average British blackbird at two months old has an expectation of further life of only just over eighteen months. About forty per cent of the adults die each year, and less than ten per cent of those fledged reach their fifth year, under one per cent surviving to the age of 9. These figures require that on average each pair should produce three fledged young every season, to keep numbers even without any net emigration or immigration.

Blackbirds are about ten inches long and weigh about three to four ounces or roughly a hundred grammes; they are therefore about the same size as fieldfares but shorter-winged and a trifle shorter-tailed. The plumage of the male is entirely glossy jet-black, the bill being orange-yellow and the legs and feet dark brown, and there is a noticeable yellow eye-rim. The female is more or less dark umber brown above and a rather redder brown beneath, with a pale throat and more or less distinct dark thrushlike spots. The young are a good deal lighter and are spotted above as well as below. More or less white or pied birds are not infrequent, and sometimes several are seen in one area; occasionally blackbirds are met with the face more or less bare of feathers.

ROBIN

(Pl. 30, xxv, pps. 145, 176)

Unlike the house-sparrow, the starling and the rook, the robin would probably remain a very common bird in the British Isles, even if the entire human population were destroyed, which is no longer such a remote contingency as it used to seem. In fact there is at the moment a more assured future for robins here than there is for us. Nevertheless robins have learnt to live with us to good purpose, and their interests would undoubtedly be injured by our abrupt extermination. Robins are birds of low cover abutting on patches of open ground, and in natural conditions they are therefore largely confined to open woodland with plenty of undergrowth and clearings, and to thickets and woodland fringes or sites created by such accidents as landslides, floods and fires. To these have been added by human action a vast range of gardens, orchards, parks, cemeteries, roadsides and waste patches or strips along railways, canals and reservoir embankments, allotments, hedgerows, windbreaks and plantations for amenity or use, while in addition woodlands themselves have been cut up or managed in ways which increase their scope for providing robin territories. On the whole the more we interfere with the landscape the more we help robins to increase. The growing tendencies to grub up hedges, reclaim scrub and thicket for agriculture and to trim hedges drastically during the breeding season, are undoubtedly detrimental, but the drastic felling of woodlands which are left to be replaced by low scrub and the afforestation of open country with dense masses of low trees must at present fully offset any such losses of robin territory. Robins in Great Britain are neither shot or trapped for food nor kept as cage-birds, and from a robin's standpoint the British people seem to have only two serious faults—their addiction to rushing through robin territories in heavy vehicles without warning, and their inexplicable passion for deliberately infesting their neighbourhoods with cats.

Robins, like dunnocks, restrict their feeding in the open to areas within immediate reach of cover, and although they perch freely on exposed posts, wires, walls and buildings, as well as on bare ground and rocks, they are not normally seen far from either bushes or rank herbage such as nettles, bracken, reedbeds or suitable crops. They not infrequently enter buildings, especially conservatories, and thirty years ago at Sedbergh one used to enter through the open skylight of Hart House grub-room and pick up the crumbs dropped by hungry schoolboys devouring cake.

Robins perch more freely in trees, especially for singing, and show more preference for their neighbourhood than either dunnocks or wrens, which is natural since they are apparently by origin birds of forest undergrowth and fringes, and in many parts of the Continent are still wholly or mainly restricted to woodlands, while the more open and artificial types of habitats they would occupy in Britain are often dominated by redstarts, black redstarts, or nightingales. In the Azores, on the other hand, it appeared to me that the robins showed themselves more in the open than in England. In southern Germany and France, and in Switzerland and elsewhere, robins are common in mountain forests up to five or six thousand feet or more, but in my own experience in the British Isles, while robins flourish up to 800 or 900 feet they are shy of going higher. On suitable coasts they will forage and even sing on the beach, if bushes are near.

Robins breed right up to the North of Scotland, and in the Outer Hebrides, which they have only colonised within living memory. They seemed to me as common at Stornoway as in a normal English district. They are, however, very patchily distributed in the Hebrides and the northern Highlands, choosing only the most favourable sheltered areas, and have not yet reached Shetland. In Ireland also numbers fall off in the more exposed districts of the Atlantic coast. Many leave Scottish and northern districts for the winter, and some cross the Channel to the Continent, British-ringed robins having been recovered in France and Holland. At suitable points such as Spurn in Yorkshire the autumn passage of robins can be clearly observed and the passing birds caught and ringed.

Robins have so thoroughly mastered the art of unsociability that even to see as many as four adults close together is an uncommon sight anywhere except on Christmas cards. Wherever there are two robins of either sex living close together, unless they are actually breeding, they find it hard to restrain themselves from holding territory against one another, and two passage birds have even been recorded as doing so at opposite ends of a ship on which they were resting at sea.

Robins are the classic demonstrators of territory among birds, and their territorial reactions and habits have recently been thoroughly described and analysed by David Lack in his *Life of the Robin*. M. Brooks-King has described how a young robin, after losing most of its head-feathers in autumn territorial fighting, retreated into a house and established territory both upstairs and down, having at first to be protected from attack by keeping the windows shut against hostile rivals in the garden. Robin territories take shape immediately after the summer moult, when the intense outburst of autumn song proclaims a period of great activity in which not only the cocks but about half the hens sing songs with no distinguishable difference, according to Lack. Both cocks and in many cases hens at this time seize individual territories averaging when observed in South Devon about ¾ acre each, but with wide variations. In the course of acquiring and vindicating these territories much fighting takes place in August and September, and to a less extent later. Song, however, is not an invariable accompaniment: a robin which took up autumn territory about my house in Chelsea during October-November 1949, was only heard to sing a single brief snatch in several weeks. Pairs are formed from mid-December onwards on the initiative of the hens, who sometimes mate with the cock on an adjoining territory but often move over to a new place where they posture and sometimes sing loudly at each other, with intervals of taking food together from the ground until the cock accepts and follows closely after the hen and the pair is formed. After this the cock tolerates the hen's presence and can recognize her at least thirty yards away, while the hen learns the boundaries of her mate's territory by dint of being chased back over them by the neighbours. Birds paired early spend several weeks almost ignoring each other before the hen single-handed builds the nest, usually in late March or April, when coition follows on the hen holding herself still in a characteristic attitude. Other robins, and occasionally also dunnocks and other birds, are chased out of the territory or urgently invited to leave by the threat display, which consists of puffing out the red breast and exposing it to the utmost against the intruder with a swaying or turning movement and the neck outstretched. If the intruder does not leave fierce fighting may follow, and in rare cases the loser is actually killed. Most often, however, the loud warning song reinforced by occasional pursuits is sufficient to clear the territory, although trespassing is frequent. The largest stable breeding territory surveyed by Lack in Devon was about 2 acres in extent and the smallest about two-fifths of an acre with an average of 1.4 acres, while Burkitt at Enniskillen found an average of 1.5 acres, and I obtained a similar density on a Surrey estate after deducting open ground. In some cases territories may consist of two or more distinct areas separated by open or otherwise unsuitable ground. About June territory-holding disintegrates in robins as in most other small territorial species, and considerable influxes of strangers may be detected in some places, although the onset of the moult makes robins exceptionally secretive and so hard to find that it is almost impossible to gauge their status in areas visited in or around July. The old story that the adult cock drives off or fights the young is groundless, robins of any age at this season being too dispirited to fight anything, while after the moult the point does not arise since all have become strangers.

We are also indebted to Lack for having worked out the mortality of robins. While there are records of wild robins living as long as nine, ten, or even eleven years, very few attain more than a fraction of this life-span. About one-fifth of cock robins are unmated, and each hundred birds at the beginning of the breeding season therefore include about 45 breeding pairs and 10 unmated cocks. Laying usually 5–6 eggs and rearing two broods each pair is estimated to produce an average of six fledgelings each season, but many of these die very young, and by the beginning of August ringing returns suggest that about a quarter have already perished, while of those still living on 1st August it is estimated that 72 per cent die within a year. Adults alive on 1st August show little higher survival rate, about 62 per cent being doomed to perish before a year has passed. Thus assuming a stationary population almost exactly two-thirds of the robins living at 1 August each year will be juveniles and one-third adults. Although robins are quite capable of living over ten years the wild robin of three or more years old is in practice a rarity. As Lack says, who killed cock robin is for the most part still a mystery. Cats, mousetraps, cars, birds of prey, hard weather, drowning and many kinds of fatal accidents are known to take toll but the great majority die unnoticed and from untraceable causes. With a pulse-rate of nearly a thousand beats a minute, or fourteen times our own, and with a habit of adding to the inevitable dangers of life by going about looking for trouble, robins no doubt manage to burn themselves up or to meet sudden doom even more often than observation would suggest.

One conspicuous risk which robins take is in hastening to investigate closely anything new or strange which they meet. I have seen a robin on the way to feed a brood with a beakful of insects turn aside on hearing the cry of a wryneck, and flit up to the top of an oak to quizz the stranger from a perch within a couple of feet of him. Another singing on a fence broke off to drop to the ground for a ringside view of a fierce fight between two cock chaffinches which came down locked together in a duel. Another, at Cookham Dean, the moment a baker's boy left his horse and cart outside a drive gate, dashed out of the ivy on a wall, settled on the middle of the horse's back, hopped up on to the cart and pecked off a large crumb with which it darted back into the hedge, returning for another crumb before the baker reappeared. In all these cases the robin came to no harm, but this boldness makes the cocks especially among the easiest of all birds to trap, and many must rush into situations from which there is no escape. Their reaction to recognised danger is however extraordinarily quick; one which I was watching in Surrey, uttering a very sweet continuous sub-song from an apple tree, reacted so fast to the sudden discharge of a gun within a hundred yards, that his form looked quite blurred as he instantly dashed for cover. Another at the same place spotted a stoat hunting through his territory and came down within a yard continuously making the "fisherman's reel" alarm, which was taken up by two or three neighbour robins who thus had good warning of the invader.

Robins carry themselves with an engaging boldness, perching with legs firmly planted wide apart, body upright and red breast thrust confidently forward, the head being cocked alertly to bring to bear one or other of the bright beady eyes. On the ground they hop rapidly with frequent thrush-like pauses. The flight is darting and determined on brief sorties, but flitting and usually low on longer movements.

Of the vast sum of bird songs uttered each year in the British Isles, robins must contribute more than any other species. Probably the only more plentiful birds are the chaffinch and blackbird, both of which hardly sing at all during six months of the year, and the starling, whose song is less interrupted but much less in volume. Robin song attains by about the end of August an intensity comparable with that of spring and its quality at this season is perhaps even better than in spring; from then on it continues to be a frequent everyday sound until well into June, the only weeks when it is not to be heard being during July. In autumn morning song is much more frequent than evening, but in March and April the robin, like the blackbird, is more fond of singing in the evening. Even darkness does not silence it; I have heard plenty singing in the rain in January forty minutes before sunrise.

In pattern the song is a series of brief snatches of pure, sweet and fairly loud notes fused perfectly together, sometimes shrill but never harsh, and of no great carrying power. In pitch the spring song has about the same range as a dunnock's, from 1,400 to 4,500 vibrations a second, but Ludwig Koch has recently proved that the autumn song contains passages of much higher frequency, which are well beyond the capacity of the human ear.

The robin's alarm has a metallic timbre and a mechanical tempo reminiscent of the slow paying out of a fisherman's reel, the speed being increased under stress of excitement; there are several other notes.

Robins eat mainly insects, especially beetles, moths, ants and flies, but also spiders, worms, seeds, soft fruit and berries. There is a record of a wild robin, which had been encouraged to come into a house, living largely on butter and margarine, and even feeding his mate and young on it, while another bird, an old cock, when presented with butter and margarine always helped himself to butter, which he could distinguish before having tasted it.

Although the nest is often made under the eyes of human beings it is built and visited with the greatest secrecy, and is often difficult to find even after the young have been hatched. One nesting in a hedge-bank near Selborne would dive off when disturbed and run mouse-like along a half-tunnel under the roots, taking refuge in the deepest recess and refusing to be driven out of it. The hen builds the nest in a bank, in ivy, in a hedge-bottom, on a ledge in outbuildings, or in some hole, such as an old kettle, a nesting-box, a hollow stump, a coat pocket or even a letter-box. The nest is usually of dead leaves and moss lined with hair and is big for the size of the bird; there are normally 5 or 6 eggs and two or sometimes three broods are reared. Building sometimes starts in March but often not till well on in April. Incubation and fledging each take about a fortnight.

The dark olive-brown upper-parts are set off in both sexes by a large bright reddish-orange patch, reaching from the forehead down the breast, fringed with bluish-grey and succeeded by white on the belly. The young, until about August, are without the red breast and are coarsely spotted and spangled with buffish markings both above and below—a plumage confusing to those who imagine that there cannot be a robin without a red breast. Robins are about 5½ inches long, about the same size as a whitethroat but with slightly shorter tails and slightly longer bills and legs. They weigh about threequarters of an ounce or 20 grammes.

DUNNOCK

(Pl. 31, p. 160)

Dunnocks do no harm to us, but have in return been exposed to the undeserved insult and injury of being miscalled hedge-sparrows by people too stupid to see the absurdity of such a name, or too timid and conventional to revert to the older, briefer and better one. It is true that the misleading "hedge-sparrow" has the authority of various eminent authors, including Gilbert White, but we do not follow White in calling reed-buntings reed-sparrows, which would be much less unreasonable. The perpetuation of the unfortunate "hedge-sparrow" is not merely scientifically unjustified, but is calculated to cause destruction of these harmless birds and their nests whenever measures are taken against house-sparrows. Pedantic efforts to re-christen the species "hedge-chanter" or even "hedge-accentor" have fallen flat, and as error has less staying power than truth we may confidently look forward to the time when the whole company of British bird-watchers will call a dunnock a dunnock.

Apart from this curious historical accident of becoming saddled with a wrong name, dunnocks have done little to attract our attention considering how close to us they live. Dunnock-watching, in fact is, in my experience, one of the most baffling and unrepaying forms of bird-watching, no doubt because I have not been perceptive enough to see the interest and excitement which a dunnock's life must hold. Surely no species can maintain, as a dunnock does, a secure position among our ten commonest birds, without having exceptional qualities of some sort. Yet most of the more evident points about the dunnock are negative. Although a soft-billed species it clings to its haunts, winter and summer, more persistently than any other comparably abundant small bird except the hard-billed and tougher house-sparrow. The British and Hebridean subspecies migrate hardly at all, although there is some passage along the east and west coasts. This ability to stand its ground is based partly upon the dunnock's capacity to live in winter almost entirely on seeds of a variety of very common uncultivated plants, including such weeds as knotgrass, plantain, sheep's sorrel, chickweed, and buttercup, while in summer it switches largely to insects.

Again, the dunnock is not to be described as a bird of any such clear-cut habitat as woods or fields, but creeps in anywhere where a sufficient patch of undergrowth or low cover commands a little open space in which insects can be caught and seeds gathered. Such places are found in the more open woodlands, where the tall canopy is not too thick, and in natural thickets, clumps or bushes and scrub or even tall rank herbage, bracken or heather on heaths, moorlands, downs, marshes and shingle. Artificially, also, they are maintained on a large scale in gardens and parks, orchards, hedgerows, waste patches by roads and railways, or on industrial sites, borders of reservoirs and canals, and plantations of young trees.

Being thus a bird of fringes, margins and habitats in transition, the dunnock has flourished greatly through the modern shaping of the landscape into an affair of artificially maintained fragments of different habitats side by side, and of constant destruction of matured and balanced plant and animal communities which become an ecological no man's land. Into such humanly created wildernesses the dunnock comes readily. It is in fact a born squatter, unobtrusive, unexacting and adaptable, flourishing on shifting odds and ends of cover and of feeding-ground which would not meet the needs of most species. I have seen a dunnock singing in July inside the Temperate House at

Kew Gardens, and it is also to be found in the Zoo and many smallish open spaces of inner London as well as all the central parks. In suburban gardens it is one of the most flourishing and widespread of birds, and it is well entrenched in suitable low cover in all directions as far as the cliffs and foreshores, and also even on outlying islands. It is common in the Scillies and is missing as a breeding species only from Shetland and St. Kilda, and from some of the smallest and bleakest outliers of the British Isles. Abroad it is represented by various races almost throughout Continental Europe from the Arctic to the Mediterranean and Black Sea, but from mid-France and the upper Danube southwards, it tends to become restricted to the high forest areas, up to the tree-limit. In Tirol and also in Switzerland, I met several singing at c.6,500 feet in the conifer scrub.

Dunnocks rank high for unsociability both among themselves and with other species. The families break up in late summer, after which it is unusual to see more than two or three together, although this dispersal seems to be maintained without the incessant aggression characteristic of robins.

Longevity is not one of the main factors in the success of this species, and the evidence points to dunnocks suffering as high an average mortality and having as low an expectation of life as almost any of our small birds, although they appear less prone than some to suffer disastrous losses in severe winters. As well as undergoing all the normal hazards of small birds dunnocks are exceptionally liable to be picked upon as foster-parents by cuckoos, and there is a case on record of one unfortunate pair which had two cuckoo's eggs laid in each of two successive nests which they built one season.

Although good at moving in dense undergrowth, dunnocks have a slightly clumsy-looking gait, sometimes hopping, sometimes creeping. They usually sit well forward, with chests thrust well out, and are rather quick and nervous in their movements, wagging the tail briskly up and down from the base through a narrow arc. They like to be in or under cover, and although often foraging in small open patches or near hedges, are extremely averse to venturing out in open fields or other unprotected places. They will however sit on top of hedges or bushes, and even perch on house-tops or in well-grown trees, but rarely so as to make themselves conspicuous or to sacrifice immediate access to dense cover. They are not, however, skulking birds like many of the warblers, and can usually be located and flushed with little trouble.

All the same, owing to their habits they are among the most easily missed of our common birds, and close investigation is needed to find out at all accurately their numbers relatively to other species in an area. Fortunately they are constantly uttering call-notes which though not especially loud or striking are usually easily recognized, calling attention to the presence of many dunnocks which would otherwise not be noticed. Dunnocks' voices are of moderate volume, rarely faint but never very loud, and their principal characteristic is a combination of fairly high pitch with a rather rough or at least a piping delivery. The typical call is a high-pitched but throaty *zeet* sometimes doubled and slightly clipped, sometimes slurred into a quick "it-it-it". A cry of intense alarm and anger repeatedly uttered by a Surrey bird demonstrating against a prowling cat at close quarters was an agonised louder and shriller *Dee, Dee, Dee,* much purer than the usual notes but with a very faint ring in it; the passion behind this utterance was most striking, especially in a bird whose usual notes sound so flat.

The song is a brisk unpretentious warble lasting up to four or five seconds and typically delivered about seven or eight times a minute with just enough volume to carry about two hundred yards, but not enough to save it from becoming easily drowned in a chorus of other songs or background noises. It is neither pure in tone nor finished in delivery, nor does it even impress itself on the memory by any striking harshness or emphasis. It is sometimes given in flight. There is also a faint subsong and in addition the cock in courtship will use a low sweet warbling song not audible anything like so far as the normal. On one occasion in Surrey in April I saw this addressed to a presumed hen sitting within a yard of the singer who became tremendously excited, fluffing out her plumage so as to look twice the size of the cock, and flicking her wings and putting herself in an attitude of invitation at the same time, uttering a low emotional cry, not unlike that of a hen house-sparrow in similar circumstances. The cock also flicked his wings, but was torn between response to the hen and the distracting challenge of a rival cock's song near by; in the end he flew off without pairing.

Edmund Selous and several other observers have recorded hen dunnocks assuming a similar attitude with the effect of leading the cock to peck gently at the rump exposed by the erected feathers, thus apparently producing sexual satisfaction. Chases between two or three dunnocks are frequent, especially in spring, and are usually difficult to follow. One rainy April morning I watched my pair of dunnocks racing full pelt round a bushy carnation plant in the garden, one with a fat grub in its bill; after several times round the pursuer suddenly reversed whereupon the other stopped dead and with great presence of mind swallowed the grub on the spot.

The flight is low, fluttering, quick, and often rather twisting, and is usually confined to short distances.

Dunnocks usually nest fairly low, in bushes or creepers, sometimes on a bank or other more solid support, and the nest is on a twig foundation but has a neat cup lined with moss, hair or wool. It is built usually in April and four or five deep blue unspotted eggs are normal, and are conspicuous when the bird is not sitting, although I have found nests with the eggs lightly covered by dead hawthorn leaves in the bird's absence, and in one case I felt hard objects under some untidy wet moss in an apparently empty nest, and discovered two eggs hidden beneath it. The eggs hatch and the young fledge in about twelve days each, and two or occasionally three broods are reared.

Dunnocks are about $5\frac{3}{4}$ inches long, being longer and slenderer than robins but shorter in the wings, bill and legs. Plumage varies in detail between individuals but the upper parts are warm brown, much mottled and streaked with blackish, and the crown, nape, face, throat and breast are more or less smoky-grey, with reddish-brown patches at each side of the breast which sometimes make a partial or even a complete band across it. The belly is also greyish, and there is no striking difference on account of age, sex or season. The thin longish bill is blackish, and the legs and feet fairly bright flesh-colour. The Hebridean race is darker and the Continental paler and brighter than the British, which is thought to extend also to western France.

WREN

(Pl. 32, p. 161)

For most bird habitats we can, without undue distortion, use those large loose groupings which enable us, with admirable assurance, to dismiss the most astonishing complexes of plants, animals, and miniature climates as, say, "woods" or "heaths" or "marshes". A wren's world, however, is more comparable in some ways to a mouse's than to our own, and the wren cannot adequately be described as a bird of woodlands, gardens, fields, moors, marshes, cliffs or wastelands—

although it is all of these—but must be looked at rather as a bird of crevices and crannies, of stems and twigs and branches, of woodpiles and fallen trees, of hedge-bottoms and banks, walls and boulders, wherever these may occur. Wrens, therefore cut across, or rather scramble under, the imaginary boundaries which we are accustomed to draw between different types of country, and each of our familiar habitat types may hold many wrens or few or none, according to its richness or poverty in these often overlooked and despised amenities.

Tiny in size, the wren can slip in and out of almost anywhere, and needs no great bulk of food to sustain it. Vigorous and active, it can scrape a living wherever moths, flies, beetles, aphides, spiders and other such small edible creatures are to be found accessibly, in their various stages of growth. Sturdy and mobile, it maintains itself throughout even the more exposed parts of the British Isles, being represented by separate subspecies in St. Kilda, Shetland and the Hebrides, although absent from North Rona. On the Atlantic coast of County Clare I have found a wren inhabiting three low stunted trailing brambles forming all the cover in sight in a place so bare that even turf was less in evidence than naked limestone. In the Dungeness area I have found one in the remotest outpost tufts of furze on shingle where not a blade of grass grows. It flourishes on spray-swept slopes under the cliffs, even visiting boulders on the beach with the rock-pipits, while in highland areas it ascends freely to at least 1,300 or 1,400 ft., in Snowdonia regularly to 3,000 ft. (B. Campbell), becoming a neighbour of the red grouse and golden plover, and in the East Anglian marshes it may be found at home with bitterns and bearded tits in the big reedbeds. On St. Kilda the local subspecies is to be seen on Conachair, the greatest precipice in Britain, and it nests among the fulmars and Leach's petrels. With the exception of the barest slopes and peaks of our highland regions and the most densely built sections of our large towns it is doubtful whether there is any significant patch of cover for living animals in the British Isles which is not at least occasionally inspected by wrens. They do not however manage to breed in inner London as regularly and generally as robins and dunnocks, except in Battersea Park, which appears to be the only open space providing favourable nesting situations. With all their enterprise and restless curiosity, wrens rarely venture far into open fields or other places more or less devoid of cover, nor do they often explore woodland canopies or the upper parts of trees of any considerable height, although

H. M. Stone

a. Dunnock with food

Robert Atkinson

b. Dunnock's nest and eggs

PLATE 32

C. W. Newberry

Wren's nest in broccoli

they are fond of searching the lower trunks and boughs. They like creeper-clad houses and all kinds of outbuildings and greenhouses, the more ramshackle the better, but they avoid the more sterile and taller types of buildings, and they have taken little advantage of many apparently favourable openings created by bomb damage in town centres. Possibly the accumulation of animal and vegetable debris from past years in such places, has not yet had time to reach the stage really attractive to a wren's eye.

Although perhaps the most widespread bird in the British Isles, and in certain areas and at certain times amongst the most numerous, the wren is undoubtedly far outnumbered in total by such species as the chaffinch and blackbird, and after a severe winter is so reduced that it must fall well below the first dozen species in order of abundance. Abroad, it is represented by one race or another from Iceland, Faeroe and Arctic Europe down to north-west Africa and the north shores of the Mediterranean, and thence eastwards in a narrowing belt through Asia Minor, North Persia and Russia to Japan, the Kuriles, and across North America—a range which gives additional proof of the stoutness and adaptability of the species. In the more southern parts of its range it is mainly a mountain bird. In France it becomes distinctly scarce in the southern half, and in Switzerland and Tirol I have noticed it in the upper forests up to about 6,500 ft.; in Bavaria I heard plenty singing just under 3,000 ft.

Wrens are not particularly sociable, either among themselves or with other birds, but occasionally give some impression of associating purposely with a mixed flock of tits and other small woodland species.

According to studies made of a wren population in Holland by Kluyver and others, the cock wren holds territory all the year round and builds in it early in spring a series of unlined "cock's nests" to which he attracts a female by song and display; she chooses one of the nests, lines it, and lays and incubates her eggs while the cock goes on building more nests and tries to attract additional females, as often as not securing one more mate and sometimes two. The cock often leaves the hens to feed the young in the nest singlehanded, but accompanies them after they are fledged, when a hedge or woodland corner will often seem alive with the tiny plump birds and the insistent, loud, very plaintive call which is reserved for these occasions.

Territories are small and compact and the extraordinary loudness and vigour of the song enables neighbouring cocks to alternate with and answer one another in a pattern of counter-singing, as E. A. Armstrong has called it. The song is a clear shrill warble of unusually high pitch which bursts out as if the singer could no longer contain it, yet is uttered with perfect enunciation and finish, usually including a high trill. It lasts about five seconds normally, and is uttered from four to six or, on occasion, as many as twelve times a minute. It carries a remarkable distance in good condition, and although at its height from March to early June, is the only song which is to be heard any day of the year in Britain. It is uttered from any convenient singing post, not often much more than fifteen or twenty feet above the ground, and quite frequently is continued on the wing between perches. It has many variants, and in addition to a feeble warbling subsong wrens, like dunnocks, will address to the hen during courtship a low, sweet version of the normal song, while according to Armstrong some hens themselves occasionally sing.

The commonest note, heard freely where wrens are about, is a moderately fast, clicking, emphatic "tititit" which may exceptionally be prolonged for as long as a minute or more. It is harder and less deliberate than the ticking note of a robin, from which the practised observer can distinguish it without difficulty. The alarm call is a loud scolding churr of "r" notes so rapidly uttered as to run together in a growling trill. Once about Christmas by Regent's Park lake, I heard a wren constantly answering with this alarm a robin's ticking call. In summer another note often causes temporary mystification, this being the loud shrill plaintive hunger-call of fledgelings or young in the nest.

Wrens have a more serious problem than most birds in keeping their tiny bodies at the necessary high temperatures through the long cold winter nights, and although they often roost singly or in pairs, larger groups are found huddled together in cold weather in holes in haystacks, or in trees, or in other cover including old nests. Dunsheath and Doncaster found nine or more sleeping heavily on a freezing February night in an old throstle's nest under eaves, all roosting with the tails pointing upwards and outwards. Ten or more have also been found roosting in a coconut shell and thirty or more in a group of martin's nests.

Although wrens are found throughout the British Isles at all seasons there are considerable southward movements in autumn, which are most conspicuous in areas least favourable to them, such as inner London and the barer tracts along the coast. Some emigrate to the Continent, return movements occurring between March and May.

Wrens have a characteristic nimble and bustling way of moving, rarely keeping still, and frequently cocking up the broad short tail. On the ground they hop, run or creep, and they can climb vertical treetrunks. They fly strongly for brief distances with whirring wings, but are rarely seen to fly high or far. Being very buoyant they can swim if they choose, and one was once seen taking pleasure in making repeated short swims across a deep waterbutt.

Display involves spreading the wings and fluffing out the feathers; I watched one cock, just after displaying to a presumed hen, attack and drive off a cock blackcap. Wrens are unfailingly brave and aggressive, especially in defence of their nests and broods, although they readily desert eggs if disturbed.

The mystery why so small a bird should build several large and elaborate nests, carefully matched with their surroundings and neatly domed, has been plausibly related to its territorial and bigamous way of life, which requires the cock to attract hens by providing the most suitable accommodation, and demands extra protection for the eggs which the hen is left to incubate alone, and which would otherwise be exposed whenever she went off the nest.

Although so skilfully constructed of dead leaves or grass or moss or other material matching its surroundings, the nest with its neat round entrance and curving dome is not difficult to find, the worst problem being to distinguish the many unlined cocks' nests from the few which have been adopted by hens and lined, and to make an inspection of the small but deep inside cup without causing desertion, especially as the hen will often sit tight even when a finger is inserted. The nest is oftenest against some solid support and background such as a bank, a wall, ivied tree, upturned treetrunk, stack or outbuilding, or in a rocky cleft, but quite commonly also in bushes, hedges or old nests of other birds. In Battersea Park I have seen a nest full of young 12 ft. up in the canopy of a hawthorn tree directly over a main pathway in a catproof, ratproof and boyproof tangle of thorns admirably chosen for survival, but normally the nest is very much nearer the ground. The white eggs, spotted with brownish-red, usually number 5 or 6 although 10 and more have been recorded; near Sedbergh I found one with seven and one with eight within the same week. Sometimes only 3 or 4 are laid. Incubation takes about a fortnight and fledging rather longer, the cock coming in only about the end of the process, although he is active in feeding the fledged young. Two broods are normal.

Apart from goldcrests, firecrests and long-tailed tits, wrens are the smallest British birds, weighing under 10 grammes and being only 3¾ inches long, of which little over an inch is tail, although the bill is almost as long as a robin's. The upperparts are warm russet-brown, boldly barred, and the underparts buff with more or less greyish or whitish and lighter barring; there is a broad creamy eyestripe. The Hebridean race is quite indistinguishable in the field, but the Shetland form is distinctly darker and larger and the St. Kilda wren greyer, more barred and slightly larger still. The plump, deep-bodied stumpy shape and the very small size are alone enough to distinguish wrens from all other British birds.

WRYNECK

(Pl. 33, p. 164)

A sad decline, which we are unable to explain or check, has pulled down the wryneck in England during the last hundred years, from the status of a familiar herald of spring to one of the least-known of our regular breeding birds. Changes in the environment throughout the period of decline have been favourable rather than unfavourable to a bird fond of parks, orchards and large gardens, dependent on a good sprinkling of rotten or diseased broad-leaved trees, and not in the least averse to living among people. Wrynecks have been little shot and their eggs little taken compared with most species. They have few natural enemies, and are well endowed with both direct and subtle ways of looking after themselves and their eggs and young. As summer migrants they never have to face severe winters. It is difficult to find any clear reason why wrynecks, instead of diminishing, should not have increased in the past century.

A hundred years ago in Sussex the wryneck was known as the rinding bird and was eagerly awaited to give the signal that sap had risen enough for a start to be made in the woods. It still visited Kensington Gardens up to 1850, and was a familiar bird of outer London suburbs until the present century. Although never known to have been established in Scotland or Ireland, and rare in most of north and west Wales, it certainly bred in Cumberland, Westmorland and Lancashire, from which it has now entirely withdrawn. At the present time it is

Eric Hosking

Wryneck with tonguc out

PLATE 34

H. M. Stone

Pair of jackdaws and young in belfry

a scarce and local breeding species in England south of the Trent, missing out Devon and Cornwall, but there are several whole counties even within these limits from which the wryneck has virtually vanished, such as Worcestershire (where it was locally common till the end of last century), and Warwickshire. In Hertfordshire where again it was fairly plentiful before 1900 only odd nests are now recorded and in some years none. In Bedfordshire one seen in June–July 1947 was described as the first reported in the county for many years. Probably the main remnants of the breeding strength of the wryneck in England are in Kent, in the Henley-Maidenhead district of the Thames valley and on or near the North Downs in Surrey. Even here to find half-a-dozen pairs within a few miles in a season's intensive search is an outstanding success. Sometimes there appears to be an encouraging increase in one area or another, but over a period the numbers merely fluctuate around a very low average, while some apparent extensions of range are simply due to more intensive search for a species whose presence may readily be missed even by the best observer who is not continuously on the watch in the right places and at the right times. Often birds prospect breeding sites without settling down, and nesting seems to be oddly casual and sometimes belated when it occurs at all. Passage migrants are more numerous and often more noisy and conspicuous than the summer residents even where these still cling on, and in many districts the wryneck has become a bird of passage only.

No sudden catastrophe has overtaken the English wrynecks. Decade by decade their numbers have thinned, their breeding range has contracted towards the south-east, and widening gaps have divided their remaining summer haunts, until only a precarious bridge-head remains north of the Channel. Once plentiful enough to be widely known as the "Cuckoo's Mate" the wryneck is in some danger of ceasing to be a regular breeding species. In the absence of any sure explanation of its decline, the least improbable hypothesis seems to be that some slight worsening in climatic conditions has been undermining its position with us. Wrynecks of various subspecies breed right across the Old World from Japan and East Siberia and China, through Russia and the Baltic to Great Britain, and as far south as the Crimea, Asia Minor, the Balkans, Italy and Algeria, while they winter in tropical Africa and India. Our own breeding stock are therefore the westernmost representatives of the westernmost subspecies. The fact that they have never established themselves in south-west England or in Wales suggests that a relatively slight eastward shift of some adverse condition might be enough to make their position in the rest of England equally difficult. In view of their highly specialised requirements—they live almost entirely on ants with some beetles, butterflies and moths—it would be quite possible for them to be badly hit by some change which might leave unaffected other species of south-easterly distribution in England, such as the nightingale. Until more research has been done we can only guess at the nature and even at the existence of such a factor, and we cannot forecast whether the trend is likely to continue, or to halt, or to be reversed.

Wrynecks feed a good deal on the ground when after ants, but they are primarily arboreal birds, sometimes clinging to trunks like a woodpecker, but more often taking to the smaller branches, on which they will perch either along the branch in the picarian manner or crosswise like a passerine. The flight is fairly strong but undulating and rather finch-like; on rising from the ground they sometimes give a rather pipit-like first impression, but the size puts them distinctly outside the small bird class.

I have seen one near Selborne hop along the branches nimbly in the crown of an oak, sometimes running a few steps and once sitting on a broken-off limb with the tail stiffly right-angled downwards. One which I watched on passage in Brittany haunted a little-used railway line, no doubt finding ants easy to catch by the track. It hopped fast on the ground with almost horizontal carriage, but at each pause the pose became more erect and more nearly thrush-like. When feeding on a gentle slope it betrayed its arboreal nature by depressing its tail against the ground, although no support was needed. Sitting on a fence the pose was slightly crouching with the wings drooping loosely, like a cuckoo's. When it wanted to look behind it simply swivelled round its neck till the head faced backwards. One took cover in a thick hedge in which it sat almost hidden peering out with sly elusive movements of the head. Another, when flushed into a hedge, preened actively and also sat quite still for a long period, but once climbed nimbly up the vertical stem of a tall shrub.

The voice of the wryneck seems to vary a good deal in shrillness, loudness and carrying power. Sometimes it carries for over a quarter of a mile, while on other occasions it sounds feeble and drowsy. It is a slow, deliberate mournful "*kwoi, kwoi, kwoi, kwoi, kwoi*" creaking rather than ringing, and not unlike a barred woodpecker's or a kestrel's peevish cries. Few can safely identify it by the call without seeing the bird. Two which allowed me to approach very close on passage set up for a moment a strange croaking hullabaloo, and another passage bird in September once made a low inward chiding clamour of several notes, not unlike the more chuckling utterances of a green woodpecker. The alarm note is not unlike a garden warbler's. The most extraordinary notes are, however, the snake-like hissing when disturbed on the nest and the "jangling sixpences" cry of the young, which at first leaves one in doubt whether it is made by a bird at all, and if so what and where. It is a very miniature sound, often lasting three or four seconds, and consisting of a rapid high-pitched run of notes not unlike the middle part of the wood-warbler's song, but fainter and unvarying in tempo. It was constantly to be heard round British Headquarters at the Potsdam Conference, there being plenty of wrynecks about with their young in the villa gardens along the lakeside in mid-July, but as usual they were very secretive. Drumming like a woodpecker has been recorded. Often they are quite silent. Cornered or captured birds will put on a neck-twisting performance which accounts for the name, and will often sham dead.

Wrynecks have an unpleasant habit of failing to decide between various nesting holes, some of which are already occupied by eggs or young of tits, which they proceed with great persistence to destroy and throw out together with the nesting material, in many cases only to lose interest and repeat the process elsewhere. They take readily to nesting-boxes, but also use natural holes in trees, banks, walls or even thatch, laying their dull white eggs on the bare floor. These may number anything from 5 to 14, and are normally laid from May to July, taking about 12 days to hatch, after which the young remain in the nest for about three weeks.

The wryneck is much the same size as a cock nightingale, but slenderer and shorter-legged. The soft plumage is indescribably richly barred, spotted and vermiculated with various shades of brown, lavender-grey and buff, the variety and subtlety of the colouring being only matched by a nightjar's or a woodcock's. The nape and back show grey in contrast with the rich brown wings, and there are several broad dark bands across the long broad, rounded or square-ended tail. The throat and chin are faded creamy yellow in ground, and a fairly broad black streak runs back across the face, while another along the crown and nape widens into a peculiar very dark brown "pigtail" on the shoulders, The bill is longish, strong and straight, and the head flattish but strikingly puckered, the crown often being ruffled like a whitethroat's. The plumage of birds of both sexes at all ages is similar.

Wrynecks arrive in England between March and early May, and leave from the end of August until October.

CHAPTER 7

BIRDS OF TOWNS AND BUILDINGS

EARLIER chapters have reviewed the principal ways in which the works of men have so far affected and been exploited by birds, and have considered the species which are most characteristic of farms, gardens, orchards and hedgerows. It remains in this chapter to discuss the nature of the habitats formed by towns and by what are called built-up areas, and to notice which birds are typical of these habitats.

Considered as animal habitats, towns and built-up areas fall into a number of classes which however are rarely capable of clear-cut separation in the field. First, the extent of ground covered is important, since conditions which would be quite acceptable for a species if found in a small town within quarter or half-a-mile of open country would be entirely deterrent in the centre of a large urban cluster five or ten miles from the nearest block of fields. Rooks are the best-known example of birds whose tolerance of town conditions depends more upon how far the town stretches than on what it is like. Air pollution also, with its blighting effects on vegetation and on insect and bacterial life, has hitherto tended to increase more or less in proportion to the size of the town, although with the greater use of electricity, gas, and oil and the improved design of solid fuel appliances some of the larger towns, including London, are improving in this respect. Air pollution also means more and thicker fog, which is confusing and alarming to many birds. On the whole, therefore, the further and more cut off a section of town is from open country of some kind the less variety and generally also the less density of bird life it is likely to show. The one species which is more often found in the centre of large towns than in the inner or outer suburbs is the black redstart, but this is due to the accident that in such cases the large buildings and the blitzed sites which it favours tend to cluster in the centre.

The height, and the style and material of construction of buildings also mean differences in attractiveness to bird life. Many of our modern office and factory buildings appear wholly unsuitable to birds, and even our more ornate public buildings, both secular and ecclesiastical, harbour little except sparrows and feral pigeons, with in some cases starling roosts. Jackdaws and kestrels, which are conspicuous tower birds in most parts of Europe, are sparse in such situations in British towns. I have personally only ever seen three jackdaws in the heart of London, one of which, probably an escaped cage-bird, sat on the roof of No. 10 Downing Street uttering low hoarse chuckles while a Cabinet meeting was being held underneath, during the economic crisis of October 1949.

For the most part bird life in towns depends upon the extent, the character and the richness of vegetation and of soil and water readily accessible. Parks and large gardens not only attract birds to themselves but increase the frequency of occurrence of certain species for some distance around them. For instance, the breeding of starlings about Queen Anne's Mansions in Westminster must depend on the ready access to St. James's Park, from which food is brought to the young. Some not very frequent migrants and birds of passage, such as the grey wagtail, which pass over central London indiscriminately can more often be noted at points in the densely built-up area than other species such as the robin which are common not far away, but will not overstep their natural habitats.

The fashion of town development which covers acre after acre with bricks and mortar, steel and concrete, asphalt and tarmac, without more than an occasional patch of common earth on which plants can grow, creates an environment which is probably equally unhealthy for birds and for men, although only birds usually have the sense to avoid it. Only the semi-domestic pigeons seem capable, as long as they are fed, of flourishing indefinitely in this human desert; even the house-sparrows need some access to plant life. When we talk of town birds, therefore, we are dealing not with birds which belong to the town as field and hedgerow birds belong to the country, but rather with those birds which are best able to creep about in the interstices of foliage or herbage and of water, between the sterile crusts of building materials. The richness of bird life in towns, which is sometimes impressive, depends mainly on the extent and variety of the belts and oases of vegetation and the pools and streams of open water interlaced among the buildings, and to a less extent also on the degree of artificial feeding and of protection afforded them, including under the heading of protection the humane and civilised practice of not keeping cats.

Which are the species most adapted to overcoming the disadvantages and exploiting the advantages of life in towns? Undoubtedly the house-sparrow ranks first, although most house-sparrows live in the parks, suburbs, or small towns, villages and farm buildings rather than in metropolitan centres. Starlings are so conspicuous flocking to roost in central London that they are often assumed to be abundant town birds, but the numbers in central areas in daytime are usually moderate or small, the main feeding and nesting areas being outside. Black redstarts (also discussed in the next chapter) are now the only songsters of the City of London, their song floating down oddly from tall buildings amid the roar of traffic. They are remarkable also in fending entirely for themselves, and in being apparently in no way dependent on food provided deliberately or accidentally by people. In many towns the swift (also discussed in the next chapter) breeds in holes in buildings, although not in the centre of the largest cities, and the kestrel too is an infrequent breeding species on some tall urban buildings. In favoured areas the jackdaw, the house-martin and the swallow, all discussed in the next chapter, nest on buildings in towns where there is ready access to more open places for foraging. All these are now probably extinct as regular breeding species in the centres of great towns.

With these eight exceptions all the birds which maintain a foothold in our towns and villages are there not because of the works of man but in spite of them. These remaining species fall into two distinct groups—the birds of trees and bushes, and the water birds. Most prominent and persistent among the birds of trees and bushes are the great and blue tits, the blackbird, throstle, robin and dunnock and the woodpigeon. The carrion-crow, rook and jay, the chaffinch and greenfinch, the coal tit and mistle-thrush, and the wren and spotted flycatcher are less widely established and need rather more encouraging conditions.

Among the aquatic birds the moorhen and coot, mallard, pochard and tufted duck and the black-headed and mew gulls, together with the semi-domesticated mute swan, have made themselves most at home on urban waters, while the herring-gull and lesser and great black-backs, and locally other species such as the dabchick, kingfisher, scaup duck and common heron are also to be found within town limits. Two passerine water-loving birds, the pied and grey wagtails, are also regular in towns where there is any open water to attract them, the grey wagtail chiefly on autumn migration.

Rather surprisingly the passerine town birds, apart from the house-sparrow, rely largely on natural food gathered by their own efforts, while most of the non-passerines including the gulls, ducks, swan, moorhen, coot, and woodpigeon are more or less frequently dependent for much of their food on the deliberate offerings of human visitors, or upon floating refuse.

With the exception of refuse dumps and of sewage disposal outlets, it is remarkable how little lodgment and sustenance birds have found in the more utilitarian elements in our towns. The factories, the blocks of offices and flats, the streets and railway tracks mean little to them; it is the towers and spires, the cornices and ornamented columns, the gables and creeper-clad walls, the parks and gardens, the lakes and rivers, the lawns and roof-gardens where the birds find their foothold. The more, therefore, that we add to the amenities of our towns the more birds we may hope to attract to them. A town which is inhospitable to birds must be strongly suspected of being inhospitable also to people.

H. M. Stone

Starling feeding young

Robert Atkinson

A swallow alighting on its nest

CHAPTER 8

LIFE-HISTORIES OF BIRDS OF TOWNS AND BUILDINGS

TOWNS and buildings and inhabited places and their effects on the birds which frequent them have already been considered generally in Chapters 2 and 7. Eight of the species most typical of these habitats are now reviewed more fully, namely the :—

JACKDAW	*Corvus monedula* L.
STARLING	*Sturnus vulgaris* L.
HOUSE-SPARROW	*Passer domesticus* (L.)
PIED WAGTAIL	*Motacilla alba yarrellii* Gould
BLACK REDSTART	*Phoenicurus ochrurus* (Gmelin)
SWALLOW	*Hirundo rustica* L.
HOUSE-MARTIN	*Delichon urbica* (L.)
SWIFT	*Apus apus* (L.)

JACKDAW

(Pl. 11b, 34, pps. 17, 165)

Subtle, sociable, active and adaptable, jackdaws are more readily observed than understood as they carry on their business around us, and although Konrad Lorenz has thrown much light on their social habits we are still far from being able to interpret all that jackdaws commonly do under our eyes.

In their distribution jackdaws are sharply limited both by foraging requirements and by their narrow choice of sites for nests and roosts. Like rooks they feed very largely on the ground, but they have not become, as rooks have, virtually field birds which retire to roost and nest in trees. Jackdaws have remained crag and tree birds, making frequent and often long sorties into the fields as their daily business requires, but markedly preferring to spend their ample leisure away from the flat surface. Unlike rooks, moreover, they will forage freely under trees (although rarely in dense woodland) and on the smallest patches of herbage on cliff-slopes, mounds, garden lawns, and in some places even in village streets and along the railway tracks between platforms.

British jackdaw communities are either crag-based, structure-based or tree-based, or a blend of two or three types. Round our higher coastlines, and in inland places where natural or quarried faces or banks are exposed, they have many large breeding stations wherever suitable crevices or even rabbit-burrows are common enough. Artificial structures, especially when they are tall, and including bridges and ruins, are freely used to supplement natural crag sites, and are often preferred where both exist side by side. Holes in trees are the other main type of stronghold, and large colonies may be based on them, especially where mature parkland gives favourable foraging conditions.

A map showing the areas where there are plenty of largish holes and crevices well above ground-level, and with ready access to soil free of tall herbage would closely indicate the distribution of jackdaws in the British Isles, except in the larger exposed tracts above 1,000 ft. altitude, and in north-west Scotland and a number of the remoter islands to which they have not spread. In any particular district, therefore, the jackdaw may range from being one of the commonest birds to being entirely absent. Perhaps the most unusual status is to find it thinly and more or less evenly distributed, although even that is not unknown. Outside Britain jackdaws are found over most of Europe, spilling over into Western Asia and North Africa.

Although some Continental jackdaws appear to winter in Britain and some of our own birds to leave, jackdaws on the whole appear to remain closely attached to their haunts all the year round, unlike other field birds such as rooks, gulls and lapwings which will turn up in flocks from late summer onwards in areas far from their breeding places. No birds are more adept at exploiting mankind, but jackdaws in their fundamental customs remain firmly conservative. The small remnant of jackdaws which have hung on in a corner of Kensington Gardens, for example, continue to live much the same life as their forerunners no doubt lived when Kensington was in the country, and would apparently be little affected if London were suddenly deserted by people. Jackdaws know how to use us without becoming dependent on us.

Over Great Britain as a whole the jackdaw may well rank among the first twenty species in order of numbers, and in certain areas it probably outnumbers all others. Hardy enough to survive all but the severest winters unscathed, more powerful than any competing hole-nester and almost without effective enemies in most of its haunts, the jackdaw has little to worry about except competition for daily food, in which it is well able to hold its own. Where jackdaws have created large communities they dominate the landscape with their constant comings and goings and their noise and business as few other birds can. Their impact on other species must be considerable, although in the one case in which the decline of another species (the chough) has been attributed to jackdaws, no good evidence seems to have been produced. In some places jackdaws have not hesitated to undertake a prolonged and often successful struggle with man, usually over nesting-sites in chimneys, roofs or belfries.

Lorenz has shown that the vital point in the formation of the highly organised jackdaw society consists in the willingness of jackdaws to share territory with others individually known to them, although not with strangers, who are attacked and driven away by the whole community. Apart from such brushes with strangers, there are frequent squabbles among breeding jackdaws, who establish a recognized "pecking order" in which each bird keeps its place by jealously persecuting those immediately below it, while ignoring those far enough below to constitute no threat to its position, and being in turn ignored by those much higher in the scale, to whom it automatically gives way without friction.

> "If a Jackdaw" says Lorenz "is beaten by a rival during the breeding season, it invariably flies to its nest and, if further pursued, utters a certain characteristic cry, which is difficult to render in letters, and which I do not propose to reproduce. On hearing this cry, all birds of the colony, first of all the mate of the one calling, come rushing up to the scene of the fight, and if the aggressor does not give way immediately he is unanimously attacked by all birds present. Usually matters do not reach this last stage. Generally the pursuing bird is sufficiently intimidated by the crowd of opponents. Often he very characteristically reacts himself in the

> same way as do all the others and joins in the chorus, uttering the same cry and going through the same ceremonial movements, which consist in ducking low down while throwing up and spreading the tail-feathers".

In doing this the bird is not trying to disguise the fact that he started it all, but is no doubt reacting like the others to the stimulus of this characteristic cry. The effect of all this is to retain the advantages of competition as between birds of nearly equal strength while giving the weak a measure of ultimate protection against the strong. More curious still is the reaction

> "when one inexperienced member of the flock loses its orientation. It is at once sought out and led home by some bird which knows the way, very often by the despot of the colony himself, who seems to have some special 'responsibility' in such cases. It is interesting to note how the bird which is about to guide the lost sheep instinctively recognizes the latter's aimless and undecided flying movements as a sign of distress and how promptly he reacts to them, in a way which reminds one of the actions of a well-trained collie-dog."

Jackdaws also will join instantly in attacking any enemy trying to carry off one of their number, old or young, so long as they see the black plumage which serves as a signal.

Jackdaws bear themselves boldly and jauntily, walking briskly on the ground with frequent diversions and a wideawake air. They are untiring prospectors, enterprising although cautious in penetrating into holes and crevices, and in trees fonder of the trunk and main branches than of the foliage and twigs of the canopy. Unlike magpies and jays they avoid thickets and bushes, and they are not fond of tall standing crops. When their breeding holes are approached jackdaws show little confidence in them as refuges and prefer to make off in a body, sometimes with much clamour, sometimes furtively, even when rooks in the crowns of the same trees may stick to their nests and remain brooding.

In trees the jackdaw is perhaps the most agile of our crows, and in flight the most skilful with the exception of the chough, not suffering from either the clumsiness and deliberation of the larger forms or the over-buoyancy and hesitation of the magpie and jay. He is the most

PLATE XXV.

Ronald Thompson

Robin alighting

PLATE XXVI.

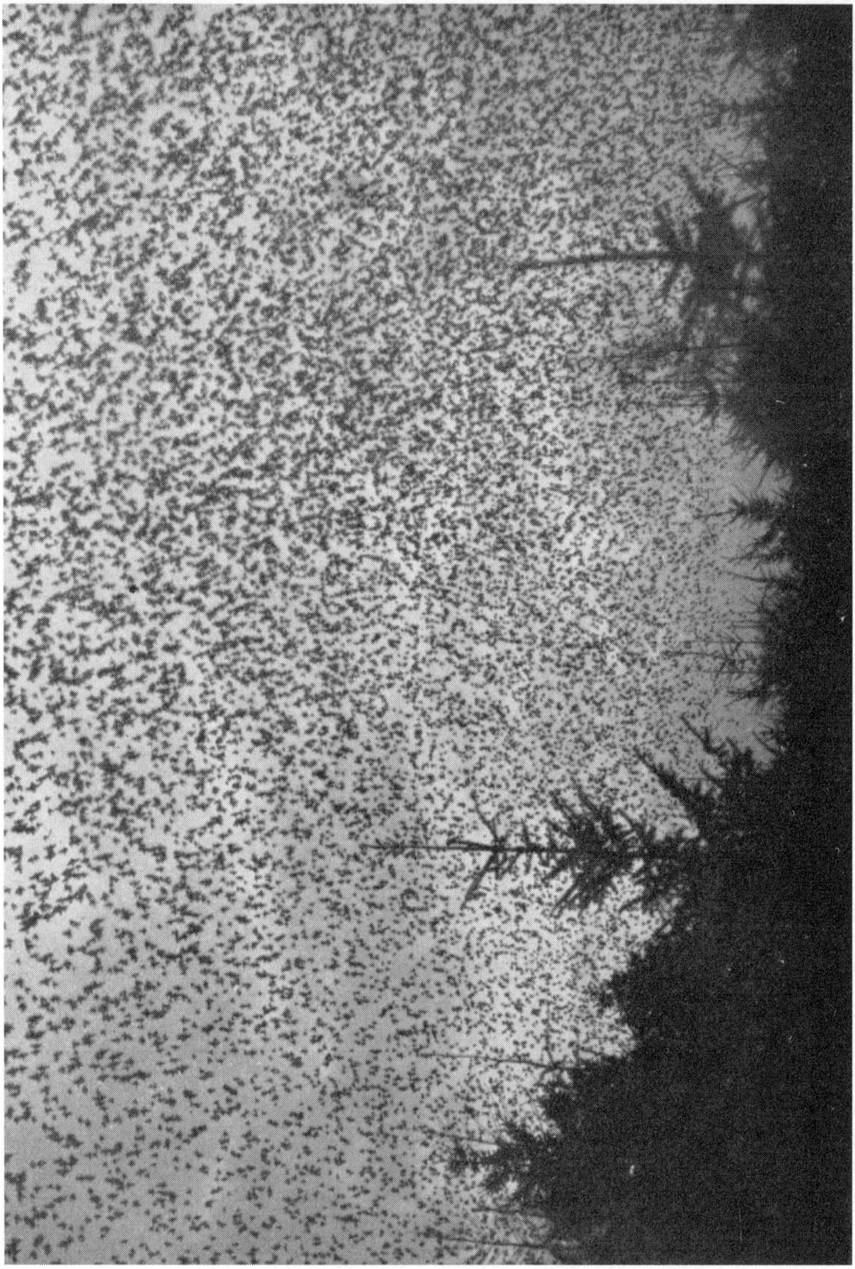

H. R. Sykes

Starlings massing above a Shropshire roost

dovelike of crows in wingbeat as in the close trim plumage and rounded head and body. Few birds spend more time or show more pleasure in communal aerial movements, for which the strong updraughts associated with many jackdaw haunts give plenty of opportunities. Although not very wide-ranging birds jackdaws habitually fly at altitudes of some hundreds, or even thousands of feet, and at an airspeed of fully 40 miles an hour. They are fond of descending at reckless speed, sometimes with rapid twisting or zigzag flight, sometimes plunging almost straight down with nearly closed wings.

Once in Pembrokeshire I saw a party of about a dozen starlings flying round overhead all making turns at the same moment followed by a much larger flock of jackdaws exactly copying them. After a time one of the leading jackdaws put on speed and definitely pursued the starlings until they exclaimed with alarm. At the same place, on a sultry August afternoon a number of black-headed gulls silently hawking for high-flying insects in a swarm about 50–300 ft. above the ground with shallow-beating wings and frequent glides, swerves and half-hovers, were accompanied and apparently imitated by at least 20 jackdaws which flopped along with deep unusually rapid beats, uttering noisy exclamations and only checking at the last instant, with fanned tails and exceptionally open beaks to snap at their prey. Shortly afterwards the performance ceased, and within an hour a cold thick driving sea-fog came in.

A well-known and peculiar feeding-habit of the jackdaw is its method of delousing sheep. On the Berkshire Downs in August I have seen as many as 6 jackdaws riding on 5 different sheep, and in the Cotswolds in November I twice saw three on a single sheep's back. In Pembrokeshire I watched jackdaws clustering thickly round a sitting sheep which had a little more fleece left from the shearing than the rest. One jackdaw flew and perched on its head and up to three others simultaneously on its back, seeming intent on getting some small objects out of its wool. The sheep remained absolutely impassive even when a jackdaw walked down between its eyes on to the tip of its nose, or clung on to its right ear with the splayed toes apparently cutting right in, or balanced with thrashing wings on the side of its face. Although there was almost a queue of daws waiting to mount this particular sheep only one other out of more than a dozen sheep received any attention from the birds, and that only from two jackdaws isolated from the main flock.

On 1 May 1918 I saw two jackdaws sitting on a donkey's back at Offington, Sussex; they were pecking, possibly collecting hairs.

Jackdaws will raid cherries heavily, and I have found some apparently gorging acorns in an oak in Devon; from their pell-mell panic on approach they seemed to be under the impression that acorns also are a jealously guarded fruit.

At a rookery in the Isle of Wight in April where jackdaws breeding in the boles outnumbered the rooks I noticed that only a minority of the daws went foraging in the fields and most concentrated on a couple of corn-stacks, where anything up to 25–35 jackdaws and 5–10 rooks could be seen all morning putting down a good beakful of wheat at each thrust, like poultry just after the bucket has been round. It seemed very doubtful whether the corn would have had much value otherwise, lying on the ground in the open eight months after harvest.

There is little that the jackdaw will not eat and its economic status is accordingly confused, but most of the complaints against it relate to its nesting-habits and its passion for picking up and hiding bright objects, rather than to its feeding-habits on the farm. Other birds, however, often view the jackdaw with suspicion, and round Selborne I found than in about half the rows between mistle-thrushes and some intruder which I investigated, the intruder turned out to be a jackdaw.

I was once watching a tawny owl in the tall beeches down the Lithe at Selborne when some jackdaws noticed him, first circling above with harsh notes of surprise and then each in turn dropping into the crown of the tree to crane their necks and hop inquisitively closer and closer, until after long and minute inspection of the still impassive owl each bird's curiosity was satisfied and they flew off one by one in silence. Shortly afterwards other birds came up and started mobbing, but the daws took no part.

Jackdaws will take young birds when they can; in Kensington Gardens I disturbed one in the act of carrying off a fledgeling starling, being warned by the clamour of the parents.

The language of the jackdaw is copious, and the species is not much given to silence. In voice, as in everything else, the jackdaw is less deliberate and ponderous than the larger crows; it has not the profundity of the raven, the sonority of the carrion and hooded crows, the gruffness of the rook, nor the hard rasp of the jay or the chuckle of the magpie. Except for the newly fledged young, who use a long drawn-out "karr-r-r-r", jackdaws' calls are emphatic, spirited, rather light in timbre and usually crisp and clear-cut. The two staple notes are a terse "tchack" sometimes two or three times rapidly repeated, and an equally loud but less penetrating "Shurr-r" sometimes heightened into a more chough-like "kyaaa". Single birds flying over will sometimes utter a loud rapid soliloquy of these notes which would do credit to a small flock. Several other notes are described, but not well, for the jackdaw's voice is even more difficult than most to render.

Jackdaws, like rooks, are fond of congregating to roost after the breeding season in trees; in south Devon I have found cliff jackdaws flighting in to roost in October in trees at the head of an estuary. It is usual for these roosts to be shared with rooks, with which jackdaws at all times tend to associate wherever both occur together. Jackdaws also associate with starlings in the fields, but do not roost with them.

Jackdaws' nests are notorious for their bulk; I once saw one in the belfry of Oxenton church in the Cotswolds which was fully eight feet high and not much less in diameter at the base. As the nest is almost always inside a hole or crevice the shape is variable, but there is usually a cup lined with hair, grass, or wool. Early one morning in Kensington Gardens I saw one busy rifling a waste-paper basket for material, and flying up to the nest with a crumpled paper bag. The greenish-blue eggs with ashy and brownish-black markings, usually four to six, are laid in April and the hen incubates them in 17–18 days, after which the young take a month or more to fledge. There is no second brood.

Adult cock and hen jackdaws are alike at all seasons, and although they look black at a distance the contrast between the soft ash-grey nape and sides of the neck and the black face, cap, and breast is conspicuous in good light at reasonably close range. All the rest of the plumage, and the bill and feet are black, but the iris pearl-grey. Young birds are conspicuously browner. With a length of only 13 inches the jackdaw is the shortest of our native crows, but this is due to its moderate length of tail and its short bill as compared with the magpie and jay, Jackdaws weigh about 7–9 ozs. (200–250 grammes), magpies weighing up to about 8½ ozs. (nearly 240 grammes).

STARLING

(Pl. 35, XXVI, XXVII and XXXII, pps. 172, 177, 192, 205)

As a successful exploiter of man's operations against nature the starling is matched only by the house-sparrow, yet we know surprisingly little about the reasons behind this success. The starling is a bird of temperate Europe and Asia. It reaches the Arctic Ocean only where Gulf Stream warmth flows in along the Norwegian coast, but has lately reached S.E. Iceland. It spreads west across the British Isles and France, with an outlying race in the Azores, but is replaced by a related species in Spain, Portugal, North Africa and the principal Mediterranean islands. It is very thin on the ground even in southern France and northern Italy, except as a winter migrant. Eastwards it extends some way into Siberia. Considering its abundance and versatility in its main strongholds this range is surprisingly limited, especially in view of the colonising strength which it has shown in North America. Since being introduced there in 1872 the starling has spread west to the Rockies, south to the Gulf of Mexico and north to the St. Lawrence, and it is quite possible that the typical race now occupies more of North America than of its original native continent. In North America also it has shown itself able to stand warmer and more extreme climates and to flourish in more southerly latitudes than it is accustomed to in Europe. Introductions have also been made in New Zealand and other Southern Hemisphere lands.

We have no definite knowledge of the past numbers of the starling, but there can be no doubt that during the last hundred and fifty years it has begun to breed for the first time in areas such as Devon, West Wales, and many parts of Scotland and Ireland, where it was formerly only a winter visitor. Over other parts of the country the breeding strength has much increased. It should not be assumed, however, that this increase is still proceeding; the peak may well have been reached, or even passed, some time since the beginning of this century. An added complication is the large winter influx of Continental breeding birds. Some of these pass through while others stay till spring, and in hard weather move from one part of the country to another. Thus it becomes hard to separate local shifts in population, varying year by year, from increasing or decreasing numbers.

Breeding densities of starlings vary enormously in different areas. Figures as high as 120 nests for each 100 acres are obtainable, although these represent concentrations of nests of birds which forage over a considerably greater area. On much typical rural land densities of between 1 and 3 per 100 acres or even less are usual. The starling probably outnumbers all but two or three of our other breeding species, when account is taken of its now general distribution throughout Great

Britain and of its strength in so many different habitats. In winter a substantial proportion of the European starling population must be quartered on us, judging by the frequency of evidence from ringing returns of visits by birds from Holland and from all the countries bordering the Baltic. It may at times become the most plentiful bird in the country at that season.

Among habitats where the starling is at home are coasts, rocky as well as muddy or sandy, estuaries, marshes and river-levels of all kinds, sewage farms, banks of rivers, lakes and reservoirs, parks and open woodlands, orchards, all kinds of cultivated land, all kinds of buildings, however large and densely packed, airfields, cemeteries and at certain seasons heaths and moorland. It would in fact be simpler to list habitats where the starling does not occur. These include some of the larger blocks of woodland and most uncultivated areas above 1,000 ft., although I have seen parties as high as 2,000 ft. on the Cheviots and other hills. That the starling is still quite capable of fending for itself without human aid is illustrated by the fact that since the people of St. Kilda were evacuated twenty years ago the starling population, which was then at a low ebb, has actually multiplied at least five times, and is now appreciably denser than on typical English farmland. This increase is probably attributable to the flourishing flock of sheep which has now taken over the main island of Hirta, including the village. The Shetland starling is a distinct subspecies, and the typical race has been much longer established on some of the remoter Scottish and Irish islands than over most of the mainland. These origins may help to explain the evolution of the toughness which has made the starling one of the most successful species in the modern world.

Starlings are very sociable, but each bird is well able to fend for itself separately. The species thus has the advantage of being able to scatter individually, or to form small or large parties, or to roll up at will into huge assemblies of as many as 50,000 or more. The gathering of these huge assemblies is made possible only by the high density of starling population and by the strong and rapid flight which enables all the birds within up to about a twenty-mile radius to concentrate at a central point within little over half-an-hour.

Tracking down starling roosts is one of the most absorbing forms of bird-watching. One Christmas, going down by train to Pembrokeshire, I noticed how the frequency and size of the parties of starlings increased all the way from London to West Wales. Near Haverfordwest

I saw a flock pass high over flying almost due north. Round the St. David's peninsula they were common right to the cliff edge and even flying over the rough sea across rocky coves. Watching on Christmas afternoon I marked the birds near St. David's setting off about 4.15 for their roost, and by projecting on a map their direction of flight I found that it intersected at a wood nearly sixteen miles away the line I had obtained near Haverfordwest. Up early the following morning I was in time to see parties coming in from that same direction to the village where I was staying. I went over to the Prescelly Hills, which were in cloud from 800 ft., and was surprised to find parties of half-a-dozen to a dozen and in one case thirty starlings busy feeding on typical grouse-moor between about 900 and 1,400 ft. up. Some were following sheep, horses or cattle, while others were in heather a foot deep, remote from habitation or livestock. Coming down below the cloud I contrived to reach towards sunset the 700 ft. ridge above the wood where the observed flight-lines crossed, and at five minutes past four the first party of starlings appeared and swept down to the roost in the valley beneath. For the next twenty-five minutes parties of between one and five hundred birds poured in on a wide front, and within the half-hour at a conservative estimate at least 50,000 birds had assembled. Calculation of the area from which the birds came indicated that it would have been by no means impossible for this one immense roost to muster twice that number, as against a total of at most 30,000 birds for all the famous starling roosts of central London put together.

At another roost of comparable size in a thicket on the Sussex Downs the birds on a February afternoon took an hour and forty minutes to assemble from the first-comers to the last stragglers, although the main mass came in in about a quarter of an hour. At a big Essex roost I have seen large flocks arrive within a mile or so and loiter about in trees and on the ground for ten minutes or more before coming in. At Hull I have seen flock after flock fly over the city and after some hesitation rise to cross the two-mile wide Humber to Lincolnshire.

The numbers at roost can change very rapidly at certain seasons. The roost in reeds at Slapton Ley in south Devon which contained only about 800 birds on 6 October 1932 built its numbers up to 2,000–3,000 by the 15th, 10–20,000 by the 16th and approximately 25,000 two days later, although no increase was visible during this period in the local daytime starling population, which consisted of only a few hundred birds within some miles of the roost. This roost, which had then been in existence at least 135 years was so noisy that it could be plainly heard more than 600 yards away.

The Winter Starling Roost survey of Great Britain directed by B. J. Marples in 1932–33 brought to light 224 then occupied roosts in England, 22 in Wales and 39 in Scotland, all roosts estimated to contain less than 500 birds being excluded. Very few roosts were found above 600 ft. and the highest was at 1,200 ft. in Yorkshire (I have seen a summer roost of several thousand birds in Durham near the source of the Tees at about 1,300 ft.). The favourite sites found by the survey were in reedbeds or trees, followed by thickets, buildings and cliffs, but in five cases oil condensers were used, the birds roosting in thousands between the warm pipes. Once when the condensers were allowed to go cold the starlings left. Although in Washington and other large American cities starlings flock to roost on large public buildings, this habit (which only started in London between about 1910 and 1920) has developed in very few other cities in the British Isles. Most roosts unless severely disturbed are reoccupied every year, and it is a serious undertaking to induce starlings to quit a roost which they have once established. At one place a fortnight's shooting with twelve guns failed, and at another £30 worth of cartridges and fireworks only drove them away for a few days. A large roost on shipbuilding sheds at Wallsend was broken up in 1948 by the noise of mechanical riveters fitted for the purpose, but as usual the birds resettled in smaller roosts close by. The smell of a roost is so strong that it has been stated to carry about 500 yards down-wind in damp weather, and to have enabled a fox to escape, as hounds drawing the roost were unable to pick up any scent except starlings'. In a number of cases it has been reported that trees and shrubs have been killed by the thick coating of droppings, while on the other hand elderberries and other fruits which the birds are fond of eating are apt to spring up in starling roosts. It has been suspected, although not proved, that starlings in their movements are responsible for introducing certain outbreaks of foot-and-mouth disease to this country from the Continent, and for spreading it from affected areas to new centres.

In some areas resident birds appear to cling to their individual or group roosting places throughout the year. In Chelsea for example, half-a-dozen to a dozen come in to roost singly in the iron turrets of

Albert Bridge, both winter and summer, and have done so for over twenty-three years to my knowledge, paying no regard to the big companies passing over at the same time to join the central London roosts. But although Continental immigrants may populate many of the larger roosts the fact that fully a quarter of the main roosts are occupied in summer is sufficient evidence that great numbers of our resident birds share the communal roosting habit. Soon after the young are fledged in late May or early June many of our native starlings leave their breeding quarters for some months; at one area which I had under close observation in Surrey they disappeared in June, shortly after the eight pairs present had reared 35 young between them, and returned on 17 September, being back in full strength and singing freely by the 20th. During the intervening months I only twice saw any even flying over.

In fine weather, especially in early autumn, the flocks coming in to roost or making excursions before settling down for the night carry out impressive aerial manoeuvres which have often been described, sometimes massing together in vast flying balls, sometimes pulling out into long skeins across the sky. At Halnaker roost in Sussex, I have seen the sky literally black with them and noted that the tops of trees were actually masked completely from view while the densest part of the flock swept across in front of them. At the old British Museum roost in October I saw some incoming flocks appear so high that they showed only as bunches of small specks, quickly becoming larger as they gathered impetus, and pitched headlong swerving and hurtling, with wings almost closed to the sides, at such a pace that they side-slipped violently in trying to change direction, and on flattening out shot across the angle of sight almost too fast to follow with the eye. I noted that the maximum speed attained in dive appeared fully four times as great as that of the flocks flying level overhead. The normal starling airspeed appears to be between 24 and 32 miles an hour, but birds migrating or flying to roost often attain about 38–44 miles an hour, or even more.

The idea that such birds as starlings can obtain any worth-while degree of improved protection from their enemies by assembling in these vasts roosts is difficult to accept. In fact the conspicuousness and regularity of the roosts, and the roar like a waterfall with which the occupants deafen one another makes them happy hunting-grounds for predators in search of an easy meal, and I have watched a prowling hen sparrowhawk pick off a bird from the edge of a wheeling flock at the roost with impunity. Another time over St. James's Park I saw a largish flock of starlings massed in high flight in a very close ball and found that they were being followed by a sparrowhawk which sometimes passed right through the flock, but without striking. A peregrine lived for some weeks on starlings at a Devon roost, and at another, where sparrowhawks had been exterminated by the keepers, eleven were shot in twenty-five days after a starling roost had been established. Owls also take toll. More probably the large roost has biological value, in spite of increased danger and inconvenience, because it brings the birds together in a central spot and enables quick switches to be made from less profitable to more profitable foraging grounds over a wide area. It may well also significantly reduce the timelag in embarking on necessary weather movements in hard winters. Much might still be discovered by an observer able to undertake intensive watching for a long period of an important starling roost and of the foraging area which it serves.

Starlings are not only sociable but are able to get along with very little bickering, not only among themselves but also with larger birds alongside which they forage in the fields. Gilbert White commented on this in his letter to Daines Barrington of 8 February 1772, and we still are little nearer to being able to account for it to-day.

Probably the closest attachment is to lapwings, and it is common to see starlings feeding among flocks of lapwings in the fields, or even trying to accompany them on the wing, when the difference in style and speed of flight is sometimes comically awkward. During the war in the course of air exercises on the upper Thames, a very noisy Whitley bomber flew over accompanied by a small handy Lysander reconnaissance aircraft following all its movements; as they passed they twice put up a number of lapwings which flew round accompanied by a single starling, parodying the same theme as perfectly as if they had been trained to. Another time in the Chilterns I saw a party of starlings flying round mixed with lapwings; eventually all but one of the starlings separated, leaving the last bird flying round for another two or three minutes in company with 31 lapwings. Finally it worked out to the right wing of the formation and parted as they wheeled, rising immediately to a height of some hundreds of feet and dashing off at unusual speed over a distant ridge. Late one March afternoon at St. Mary's in the Scilly Isles, I heard a starling "alarm note" and saw a party of eight get up and start towering to such a height that finally they became invisible even with field-glasses; they were heading north on a line towards Wales. Where starlings and lapwings fly in company it usually seems that the starlings are the hangers-on, but I have seen three lapwings deliberately attach themselves to a party of starlings flighting to roost down a Cheviot valley, and accompany them for a good distance.

Starlings will also readily mingle with rooks and jackdaws and redwings in fields, with oystercatchers and turnstones on the seashore, and with almost any other birds happening to be foraging in the same places. Their sociability does not stop at birds, for they are fully as fond of accompanying grazing animals, and their fondness for riding on sheeps' backs has long been well known.

It is curious that although non-breeding birds and even breeding males continue to flock to big roosts throughout the spring, starlings nevertheless appear to space out their nests as widely apart as possible. This spacing is not achieved by territory in the normal sense, for the birds on leaving the nest will fly straight to some meadow or other favoured area in which they forage indiscriminately, passing to and fro on parallel or intersecting tracks without the slightest friction. Is there any other bird so strongly social which so consistently checks any tendency towards colonial nesting?

With all this urge to associate with its own kin and with others, the starling can be highly aggressive, perhaps most of all in annexing a nesting-hole, which it will stop at nothing to seize for itself. Green and pied woodpeckers, both more powerful birds, are often dispossessed by the superior teamwork and tenacity of invading starlings. In Sussex I have seen some of the holes in a sandmartin colony taken over by starlings, but the great majority breed in holes in buildings or trees. Some curious sites are tenanted annually; at Sedbergh a pair nested deep down the barrel of a captured German gun, and at Dartington a pair have bred for many years in a crevice in a ruined wall barely a yard above the lawn.

Generally the starling goes to and from the nest with little or no attempt at concealment, relying on the inaccessibility of the site, and the percentage of young reared appears to be remarkably high and steady, providing the species with a good margin for expansion in favourable conditions and a resilient power of recovery should heavy losses be suffered. Among the few natural enemies able to take toll are jackdaws, and in Kensington Gardens I have seen one when scared by passers-by drop a nestling starling which it had just removed from a hole in a tree, despite the frenzied resistance of the parent birds.

The untidy straw nest is lined with feathers and usually contains from five to seven very pale blue eggs—an unusual tint for a hole-nester—which are laid mainly in the latter part of April, taking twelve or thirteen days to hatch. The young do not leave the nest for about three weeks, when their shrill penetrating clamour turns to a husky insistent coarse "tscheer"; one of the least pleasant bird sounds of summer.

The starling's song is like no other bird's. In range of content it is outstanding. Warbling, whistling, chattering, wheezing, gurgling, chuckling, clicking, bubbling and even popping sounds flow out in an unending stream, and the accuracy with which bird and other noises are mimicked is evidence of unusual skill. No other singer, with the possible exception of the robin, is to be heard singing freely for so many months of the year. The full song lasts from August until May, and even during June and July the volume is often considerable. Yet with all these advantages the song is feeble and disappointing. So successful a bird has had no need to develop the uses of song as a territorial challenge or signal, and it remains so subdued, primitive and formless as to compare with the subsongs rather than with the full songs of more advanced songsters. There is some reason to suppose that most if not all the higher songs were evolved by selection out of similarly varied and formless material, and it is clear that the starling could sing extremely well if it ever became biologically worth while for him to take the trouble.

The failure to develop an important biological role for the song probably accounts for the fact that it is so little seasonal, and is used at all times as a cheerful soliloquy or as a chorus in the fields or at roost. Even the discovery of a rich source of food will stimulate a party of starlings to burst forth into hurried thanksgivings as they tumble over one another with eagerness to get their share. The whistle heard in the song is also used as a separate call-note, and there is a loud excited metallic alarm of very high frequency, which always makes me look up at the sky when I hear it, as it so often signals the approach of a hawk.

On the ground starlings walk briskly with their heads jerking awkwardly all the time like moorhens; when hurrying they run or hop freely. In the air the plump blunt-ended body is propelled by strong businesslike fairly deep wing-beats, continuous or slightly interrupted. Passing flocks make a rustling audible from far below. In summer the sky is sometimes dotted with starlings circling and gliding not ungracefully at heights between 60 and 300 ft. while they collect huge beakfuls of large high-flying insects.

Because of the numbers and conspicuousness of starlings it has commonly been assumed that they must have a great influence for good or ill upon agriculture. This remains to be proved and the more we discover the more improbable it seems. Kluyver has shown that in Holland a pair of starlings fed their young from 118 to 525 times daily, bringing from 16,000 to 27,000 insects, of which leather-jackets formed 14 per cent in one year and 26 per cent in another. Yet the leather-jackets collected and fed to young by all the starlings on the area, where they were encouraged to nest exceptionally densely, amounted to only 1 per cent or less of the leather-jackets present, whose numbers increased in spite of the toll.

Starlings eat whatever is most readily available at the time among a very wide range of insects and invertebrates and seeds or fruits of wild or cultivated plants. Their power of rapid local concentration and their habit of shifting in search of food in June makes them a serious pest to growers of soft fruit, and they are especially fond not only of cherries but of figs and mulberries in the few places where these ripen in the open in England. An interesting proof of their adaptability is the readiness which they have shown in many areas in learning to feed on modern filter-beds over which four long black perforated pipes sweep round slowly and continuously spraying jets of water on the prepared surface. In June 1926 I first noticed how parties of starlings would feed on these beds while seeking to dodge a wetting from the jets, some running in front of the rotating arms, some flying up and perching upon them, and others ducking under them in the space between one jet and the next.

The starling is a rather unusual size and shape, most similar to a blackbird but with only two-thirds the length of tail and with slightly shorter legs. Its weight is about 3 ozs. or roughly 80 grammes, being only about three-quarters of a heavy blackbird's. At a distance the adults seem wholly dull black in plumage, but on a close view they are seen to be more or less brilliantly glossed and spangled with greenish, purplish, bronzy and buffish tints, varying a good deal according to the individual and the time of year. The young are much paler mouse-brown birds with dirty whitish chins. The bill is conspicuously yellow in spring, but turns dark in the late summer. Its base is bluish in the male, pinkish in the female. The legs and feet are reddish-brown.

HOUSE-SPARROW

(Pl. XXVIII, XXIX and XXXI, pps. 193, 196, 204)

Of all British birds the house-sparrow is undoubtedly the most dependent on man—so much so that it would be notable to find a single site where sparrows rely all the year round on naturally available food and shelter. The disappearance of the human inhabitants from Britain might well lead to the utter extinction of the house-sparrow. The geographical distribution of the human and sparrow populations is similar, but sparrows unlike people, reach their densest concentrations not in slums and tall built-up almost treeless areas, but in city parks and gardens, and in the docks, railway depots, allotments, cemeteries, waste lands, sewage farms and other common service sites within or surrounding towns.

Sparrows are often in considerable numbers also on farms, especially where grain is grown, and they gather in flocks round the ricks and hedges bordering fields where rich gleanings are to be counted on. But they are not birds of the fields or of open country, and do not care to go far from some fairly tall cover—buildings, ricks, trees or thickish hedges—from which they can sally forth to forage, or to which they can readily dash back when danger threatens. It is this talent for being always on the spot when there are easy pickings to be had and making himself scarce when there are none, that distinguishes the sparrow. He is a sharp-eyed, quick-witted exploiter, and almost alone among the smaller birds, relies at all seasons of the year upon the advantages of operating in gangs. A flock of over a hundred house-sparrows is unusual except in the harvest-fields or locally at urban roosts—much rarer than say a large flock of chaffinches, or linnets or skylarks. On the other hand a solitary sparrow, or pair of sparrows, at any season is something noteworthy, and usually only to be found on the fringe of their distribution. It is in bands of tens, twenties or fifties, not in ones and twos nor in thousands that sparrows like to operate.

Londoners and many others are apt to assume that the house-sparrow is the commonest British bird. While much remains to be found out about the sparrow population there is no doubt that the sparrow is heavily outnumbered over Great Britain as a whole by the chaffinch and the blackbird, and is probably also substantially exceeded by the starling and the robin. Of Britain's 56 million acres only some 3 million are built up or used for gardens, allotments and similar purposes. The whole area inhabited by house-sparrows is unlikely to exceed about 5 million acres, leaving more than 90 per cent of the land area entirely sparrowless. This may seem startling, but it is confirmed by evidence from sample censuses of typical farmland, woodland and even areas containing buildings in various parts of the country, in which the sparrow either does not occur, or is far down the list. For instance, on a forty-acre estate in Surrey carrying some 266–288 birds in spring and 258 at the end of the year, there were never less than 30 chaffinches and 20 blackbirds, but no house-sparrows at all. On an Ayrshire estate of about half the size my brother found between January and the end of May 1927 a total of from 136 to 236 birds of which house-sparrows accounted for none in January, 6 in February, 3 in March, 5 in early May and a maximum of 11 in late May, while chaffinches only once fell below 45. Summer counts on an Oxfordshire farm showed chaffinches to be twelve times and blackbirds fifteen times as numerous as house-sparrows.

Even in its greatest strongholds the numbers of the house-sparrow are much exaggerated. A serious work on the bird life of London forty years ago said "London may be described as a huge colony of House Sparrows, possibly numbering millions", and quoted Hyde Park as a place where their numbers "are literally enormous". Hyde Park is a part of the same open space as Kensington Gardens whose sparrow population was carefully counted by my brother and myself completely on two occasions and incompletely on two others, during the winter and early spring of 1925–26; the result showed a sparrow population of 2,603 in November, 2,595 in December, 1,840 in January (a drop due to special weather conditions) and again 2,505 in March. The Gardens cover 275 acres, and all the indications were that they formed a quite exceptionally large and dense concentration of house-sparrows, yet the pre-war average worked out at fewer than 10 to the acre, and the present number is much less. Even assuming the whole county of London to be as well stocked with sparrows as Kensington Gardens the total would not have reached three-quarters of a million, and allowing for the large proportion of the county which clearly carries a much lower sparrow density than Kensington Gardens, it was very difficult even twenty years ago to justify an estimate of more than, at the very most, 300,000 birds. Surveys in the spring of 1945 and 1950 have indicated a density in Battersea Park—an exceptionally favoured area—of some 5 sparrows to the acre, and the immediate post-war density in Kensington Gardens and Hyde Park was undoubtedly much lower than this. The house-sparrow population of Great Britain cannot recently have exceeded at the very most some 5 million adults, and I should regard even half that figure as being excessive for a probable, as distinct from a ceiling estimate, at the present time.

On his chosen ground the sparrow has the enormous asset in the struggle for existence of being virtually indestructible. He can be shot, poisoned, trapped, frozen, swamped, scorched and persecuted by every method known to nature or to man, without appearing any the worse for it. I have known sparrows to be successfully banished from outlying farms, but normally the game is not worth trying. Yet the species is highly vulnerable to one thing—a falling off in the amount of food to be picked up without much trouble. The steep decline of horse traffic, and of the food spilt by the horse at both ends has hit the sparrow hard. So has wartime evacuation of towns, and food rationing. Cereal acreage partly recovered during the war from its long-term decline, but the recovery is partly temporary and in any event does not appear to have caused any substantial and widespread increase in sparrows. In parts of the United States there has been a significant decline in the numbers of the "English sparrow", and this is attributed partly to the replacement of the horse by the petrol engine. It is not to be expected that so adaptable a bird will seriously lose grip, but the period of its great increases seems to have given place to one of stability and locally even of decline. In so far as cereal cultivation gives way to greater concentration on grass and livestock, as town and country planning makes towns more habitable for other small birds, and as other species learn to adapt themselves to civilisation and to catch up some of the sparrow's lead in this respect, the relative share of the house-sparrow in our total bird population may be expected gradually to fall back.

One of the sparrow's great competitive assets is not being a territorial bird. Holding of individual territories is a system best suited to birds which have to rear broods by collecting at top speed immense quantities of small insects quite close to the nest. It is not suited to birds whose food is accessible in steady abundance in the breeding season, least of all when the food supply tends to be concentrated, and not always at the same point. The sparrow does not choose to nest at all at places where there is not an easy living within range, usually for several pairs, and is thus able to save the vast amount of time and energy which other birds devote to song and territorial vigilance. By their capacity for raiding in fair-sized bands areas in which easy food is to be had, and by their thrusting and quarrelsome attitude, sparrows are often able to discourage and crowd out small territorial birds which are possible competitors. In a sparrow-infested park the robins, dunnocks, wrens, and chaffinches are noticeably patchy and seem to cling to those parts which offer them unusually good cover and are also farthest from the main sparrow nesting strongholds.

The sparrow is one of the few birds equally ready to perch on buildings, trees, bushes or herbage, but buildings are normally his base, from which he does not wander very far. Not being a migrant, except on an extremely limited scale, the sparrow does not suffer the heavy losses often involved in migration. Being so largely a town-dweller and hanger-on of farmsteads, hard weather finds him already settled at the very points offering easier food and shelter, to which other birds, such as larks and redwings, are driven only under stress of hunger and cold.

Free of the drains on time and energy represented by territory-holding, competitive song and migration, and being intelligent enough to live where living is easy, the sparrow has a different daily time-table from almost any other small bird. His great outlet is social activity of many kinds. Nesting, roosting, feeding, bathing, dusting, quarrelling and even pairing are for the sparrow naturally social occasions; wanting to be alone, or to be out of the gang, are foreign to sparrow temperament. There is, however, an exception in the immediate neighbourhood of the nest. In Chelsea I once noticed that a dead creeper on the wall about my window, which had been common ground till about the middle of April, was then suddenly claimed by a pair as private property, and the cock drove off intruders fiercely. I several times watched these two pairing. The hen would take the initiative, settling herself down on a stout horizontal branch—always the same one—and using a distinctive low shivering call, peculiar to this occasion, which served as a stimulus to the cock. Once I observed that this sound attracted and stimulated another pair from some distance, who were seen to fly into the mulberry-tree close at hand, the second hen, after

PLATE XXVII.

Starlings fighting

Photographs by C. W. Teager

Oliver G. Pike

(*a*) House-sparrows on bird-table with chaffinch

C. W. Teager

(*b*) House-sparrows eating sandwiches

tracing the cause of the sound, composing herself on a branch and uttering her own pairing note, which was a good deal shriller. The effect upon the second cock was identical; like the first one, he seemed at first indifferent, if not actually reluctant, but the low insistent note had a powerful effect upon his feelings, and eventually he glided down to tread her with fluttering wings, breaking away only to be recalled by a fresh clamour from his unsatisfied mate.

With sparrows it is the voice that is most characteristic rather than particular call-notes. The cries are innumerable and often almost indescribable, but they rarely deceive or confuse. Some may be peculiar to the London birds; at least I have heard various calls in London that I never hear in the country. One of these is a feeble less abrupt echo of the pied woodpecker's "tcheck". The full song of the sparrow begins about the first week in February, a little before the chaffinch's, and soon becomes the most widespread attempt at metropolitan bird-music. I have listened to it in New Oxford Street, at the Bank of England, in Seven Sisters Road, on Buckingham Palace, on the chairs in Kensington Gardens, in Berkeley Square, on Hampstead Heath and in many other places. Musical notes are sometimes included, but these are simply aberrations; the normal song is a continuous soliloquy of rather spluttering excited chirps, a shrill jerky flow of notes, none particularly melodious but a few quite passable. It most resembles those of the tree-sparrow and the reed-warbler. Sometimes the singer cocks his tail in an attitude of display, but the performance is apparently almost, if not quite, unconnected with the holding of territory, and this no doubt is why so tough a bird sings less loudly than the wren or even perhaps than the goldcrest. It is in fact nearer a subsong than a true song, and is accordingly delivered without any attempt to select a prominent song-post.

The sparrow is a strong but clumsy flier, never seeming to take the smallest enjoyment in flight, but using it as a means of getting from place to place with little deviation and no attempt at style. It is exceptional for a sparrow to fly any higher than is absolutely necessary to get where he wants to or to clear some obstacle, and he gives the impression of being able and anxious to go through life without ever flying for more than five or ten minutes at a stretch. In this he contrasts particularly with such sociable finches as the linnet, redpoll, greenfinch and goldfinch, which delight in ranging far and wide over the country well above tree-top level. Yet when lured to chase a butterfly or moth the sparrow displays quite a turn of speed, and I have even seen one pursue a pipistrelle bat flying in broad daylight near the Albert Memorial. When nest-building, sparrows will fly up trailing long unwieldy straws behind them like the tail of a kite.

On the ground the sparrow hops stoutly and with little grace. When sufficiently interested odd birds or even whole parties will cling to the perpendicular trunks of trees in search of food lodged in crannies of bark, climbing the less difficult parts. They are fond of bathing, and in Battersea Park I have seen a dozen at a time ingeniously using the water-lily leaves in the lily-tank which sank as the sparrows settled on them just enough to give them enough depth of water for a good splash with the wings. They are at least equally given to dusting themselves, and in places dry ground is full of little craters in which they cower down, throwing up the dust right over their backs with their wings.

They roost apparently most often in some form of vegetation, frequently on or near buildings, rather than actually in the masonry, and at times fly fair distances to social roosts which occasionally are quite large.

Over twenty years ago I traced some flylines to trees in the churchyard of All Saints Poplar, and estimated that there were about 1,000 house-sparrows in a single plane-tree, and as many more in others nearby. This, however, is near the docks, where food, and sparrows, were plentiful but there were few trees, so that some sort of movement was inevitable to secure a satisfactory roost. At the roost they were particularly lively; the din was incessant; a not unpleasant sound, not particularly loud, but shrill and far-carrying.

Occasionally in the daytime also very large assemblies are seen; once in January in Burnham Beeches I heard a pandemonium of chirping and found them by thousands in the hollies and beeches on the ground, some apparently feeding on beechmast, and others just being sociable. The harvest-fields, where they mingle with buntings and greenfinches, often bring together large gatherings also.

Sparrows eat almost anything at times, but like man they base their diet on cereals, which they gobble up in any form from the growing ear to breadcrumbs or cake. In the Hudson Sanctuary in Hyde Park I have even seen young sparrows being fed by both parents with crumbs brought from a distance, though the trees close at hand were heavy with caterpillars which the birds ignored. Not far away I found small may hawthorns swarming with thousands of caterpillars, and plenty of sparrows about, but in ten minutes only three of them visited the tree I was watching and these fed mostly, if not entirely, upon seeds or pieces of the dead flowers.

Sparrows sometimes go in for colonial nesting in trees, usually of no great height, and by their sprawling tenements remind us that they are not finches, as they used to be reckoned, but more akin to weaver-birds. More often the domed straw nest is in holes in buildings or in clinging creepers. Sometimes it is in a street-lamp or in the hand or draperies of a statue, or in a nest seized from a house-martin or sand-martin.

Normally the nest is built from a height of about ten feet upwards above the ground; the lowest I have ever seen, in West Germany, was only two feet up, and although placed in the hollowed bole of a tree was still partly domed. In keeping with sparrow character the nest is badly concealed, badly finished and untidy, but almost always cunningly placed and more than adequate for its purpose. Few small birds' nests give so little pleasure to look at, but few are so well devised for immunity against human, animal and weather risks. The site is inaccessible, the dome protective, the lining of feathers snug. It sometimes seems that the main hazard for the eggs and young is the frequency and destructiveness of quarrels between the adult sparrows. Two or three broods are reared, usually beginning in May, although occasional breeding occurs almost all the year round. The clutch is small for a passerine, usually from three to five, and the markings on the greyish-white eggs vary a lot, but are usually ashy-grey and brown.

The plumage of the birds themselves is also variable, and individuals are more often distinguishable than in almost any other small species. In contrast to the tree-sparrow the sexes are not alike, the female being dull brown above with some buff and blackish mottling and brownish-white beneath, while the adult male has more or less chocolate-chestnut from the back of the head down the mantle contrasting with whitish cheeks, grey crown and rump, white wingbar and a broad smudgy roughly spadelike patch of black on the throat and upper breast, the rest of the underparts being dirty white. The young are like the females but conspicuously shorter-tailed and paler. Town sparrows are often definitely sooty in appearance. Albino or partly albino, scalped, tailless and other freak individuals are not exceptional. Seasonal differences in plumage are fairly slight. In build and carriage the sparrow is fairly thickset, broadbilled, broad-bodied, and broad-tailed. The length is just under 6 inches, and the weight about one ounce, or roughly 30 grammes.

PIED WAGTAIL

(Pl. xxxa, p. 197)

Wagtails and pipits are specialised creatures with rather limited powers of adapting themselves to new habitats, and even the pied wagtail, which has become one of the most dependent on men of all our wild birds, is only a partial exception. It is true that there are few parts of the British Isles where it does not breed, and that it may be seen against a wide variety of landscapes. But we may easily be deceived by the way pied wagtails seek our company, and by their conspicuous habits and appearance and fondness for ranging far and wide. When bird populations are actually counted over any large area the pied wagtail is as often as not missing, and if present can rarely claim much more than 1 per cent of total numbers, except at favoured points where flocking occurs.

Presumably the pied wagtail is by origin a cliff-nesting or bank-nesting bird, ready also to nest in hollow trees and behind such shelter as ivy, and finding accordingly no trouble in using man-made crevices in walls, thatches or roofs of buildings. Its fondness for shallow water with open edges is conspicuous, and this taste is often satisfied in the neighbourhood of farms and country buildings, although less frequently in towns. Living so largely on flies and other winged insects, and disliking woodlands, the pied wagtail also finds that man and his attendant animals provide the best and most reliable source of insect food, again in the country rather than in towns. Safe nesting sites on buildings, easy access to water, and swarms of flies have all tended to make the pied wagtail a close companion of man, whom it has nevertheless not ceased to regard with nervousness and mistrust. Ecologically, then, the pied wagtail occupies the niche of a walking house-martin, and not being dependent on feeding on the wing is just able to maintain itself in Britain in winter.

It is difficult to feel entirely happy about the future prospects of the pied wagtail in Britain. Plenty of nesting sites will doubtless remain available to it, but with the rapid spread of piped water supplies and of modern hygienic cowsheds and methods of insect control, it looks as if the phase of civilisation which helped the pied wagtail to flourish may be gradually passing. Some decline in its numbers during the second

C. W. Teager

House-sparrows in hedgerow, December

M. D. England

(*a*) Pied wagtail

S. C. Porter

(*b*) Cock black redstart, London, 1948

half of this century would not therefore be surprising, although it would be widely regretted, for the pied wagtail's graceful form, striking plumage and lively dancing movements give pleasure to many besides the regular bird-watcher.

Pied wagtails are combative birds and breeding pairs are usually well spaced, but they are accustomed to making long journeys as an everyday business and it is difficult to rank them as territorial in the same sense as many small birds which occupy during the breeding season a readily traceable area of land on which they feed, nest and sing constantly. At other seasons pied wagtails are loosely sociable, but occur in parties rather than flocks, except at the large roosts which they sometimes form in reedbeds, shrubs or other types of cover. Occasionally they form big roosts in the middle of large towns; one on a tree in one of the busiest streets of Dublin was several hundreds strong, and at its peak exceeded a thousand, while another on the glass roof of a post office in Leicester also attracted some hundreds, which dropped in directly from a great height. In neither case were the birds disturbed by noise or illumination at night, and the roosts continued to draw large numbers at a time when most of our breeding population should have had nests. Another large roost, in gorse-bushes has been described from Wales, and one in a Surrey reedbed was estimated to contain several thousand birds in mid-November.

At Polzeath in north Cornwall there is a very much frequented field above the cliffs, intersected by several footpaths in constant use, which has some exceptional attraction for pied wagtails, ten or more being present in this one moderate-sized field all day long, and up to nearly thirty in the late afternoon. I found the birds shy, rising on being approached within about 15 yards yet although continually disturbed they would settle in another part of the same field, only occasionally overflowing in small numbers to one or two of the neighbouring fields. The field held no livestock and was covered with short close-bitten turf affected by sea spray. There was no indication that its insect life was of even average richness, and no explanation of this strange and conspicuous preference could be found, unless the field simply formed a traditional halting place on the way to and from some undiscovered roost. This was in September, and the same preference was seen at the same season in three successive years, with little variation.

In summer in the Cheviots, during a plague of small flies, I saw a cock pied wagtail and a young one perched on the back of a recently shorn sheep, both raising themselves on tiptoe as they snapped at flies. The cock gave most of his catch to the young bird; both walked about on the sheep's back and the sheep walked about too. Once the cock walked down the back of the sheep's neck as it grazed, and sprang off hovering to catch a fly near its head. This convenient movable feast continued for some time before the young bird flew off, followed by the parent.

Pied wagtails with their conspicuous habits and plumage are among the more vulnerable of our small birds to various kinds of predators, including the domestic cat. They are also often saddled with rearing young cuckoos. Those which stay the winter suffer heavy losses in severe weather, and it is a sad sight to see them walking on the ice of a frozen pool unable to get at their natural food. Many migrate to Spain or France, and I have seen one wintering in Paris, flying out over the Seine off the Quai du Louvre and picking up morsels from the water or snapping at a flying insect. Although this would probably be a British bird the pied wagtail is not absolutely confined as a breeding race to the British Isles, since a few nest on the nearest shores of the Continent from south-west Norway to Brittany, just as a few of the Continental race, the white wagtail, nest from time to time in Britain.

Rival cocks face one another in a jerky dancing hovering flight, in which they utter the simple rather subdued warbling song with frequent repetitions of the *chizzick* call-note, and a good deal of swallow-like and lark-like twittering. The song seems to be used principally in the course of such contests or in excited display, or occasionally as a soliloquy, but has not been developed as a territorial advertisement, which the cock already provides by his loud call-note and conspicuous habits. I have seen one singing vigorously while chasing a cuckoo in flight over one of the Tring reservoirs.

The pied wagtail walks daintily but briskly about on the ground, moving his tail up and down quite sufficiently to live up to his name, and avoiding tall or dense herbage. The flight is strong and purposeful, but very undulating, rising and falling in long graceful pitches in time with the long pauses between each series of wingbeats. The pied wagtail does not hesitate to fly at some hundreds of feet for fair distances, calling attention by the frequent far-carrying double *chizzick* call-note. Perching on trees is not infrequent, especially for resting, but moving about among the twigs and foliage is unusual. Pied wagtails are skilful flycatchers, hovering with the long tail acting as a brake and then making a final dart.

The nest is in almost any kind of hole, often well hidden, and visited furtively with an eye on possible observers. Unless taken over from some previous occupant it is built by the hen of the usual common materials, lined with feathers, hair or wool. The usual clutch of eggs is 5 or 6; they are greyish white with rather fine grey and brownish frecklings all over. Breeding begins in April, both incubation and fledging taking about a fortnight; there are two or even three broods.

The pied wagtail is highly variable in intensity and shape of markings, some adults being very black and others much greyer. The cock in spring plumage is deep rather glossy black from the crown to the rounded tip of the very long tail; the folded wings show much white, and the underparts are white except for the grey flanks and the large black "breastplate" running up to the bill and joining the black of the upperparts by a broad band on each side below the neck. The forehead is white and the white extends in a broad almost rectangular patch down the sides of the neck, broken by a large triangle of black which projects into it from the sides of the throat. The female and juvenile plumages are greyer and browner, but too complicated for detailed description here. In most but not all cases pied wagtails can be distinguished in the field from whites by their black instead of grey rumps.

BLACK REDSTART
(Pl. xxxb, p. 197)

In the present spread of the black redstart in Great Britain we are seeing one of the most rapid and interesting revolutions in a bird's relation to man, of which we have any knowledge. It is all the more interesting because it has followed a course which could be discerned and predicted even in the very elementary state of our knowledge of bird ecology and distribution trends. In fact in discussing lost and gained British breeding species in *Birds in England*, published in 1926, I wrote "It may be predicted that if this book were being written a few years later the black redstart would need to be added to the gains". The black redstart has, within the brief span of our ornithological history, performed three separate major expansions of its British range. The first seems to have taken about eighty years, from 1829 when it was first reported as a straggler, until the early twentieth century when it had established itself not only as a regular spring and autumn passage-migrant to England and Wales and some of the Scottish Isles, but also as a regular and by no means infrequent winterer in various parts of southern England. During this period one isolated case of breeding had been recorded, in Durham in 1845. The second expansion was first detected in 1909 when R. Cooke observed a nest from which young were reared in a sheep hut at Pett Level on the coast of Sussex. Birds continued to frequent this area, where another nest was found in 1923, and in the same year S. D. Herington found a nest about four miles west, near Hastings, where a pair bred also in 1924, two pairs bred in 1925, and birds were present in the breeding season for seven more years at least. On the north coast of Cornwall breeding was suspected from 1927 and proved two years later. Ancestrally the black redstart is a bird of cliffs and mountains and this second expansion began in close conformity with its earlier traditions. On the Continent, however, the species has largely deserted natural breeding sites in favour of buildings, and the same expansion of habitat quickly followed the expansion of its breeding range in Britain. From 1926 to 1941 three pairs or more of black redstarts nested on the Palace of Engineering left over from the British Empire Exhibition at Wembley, and in 1930 and 1933 pairs bred in Kent, the first on farm buildings and the second on Woolwich Arsenal. In 1927 both a cock and a hen had been seen during spring and summer about the Natural History Museum, where another hen was found in the spring of 1936. That same spring an unmated cock settled and sang for several weeks near where I was then living in Westminster, within sound of Big Ben, and another pair began to breed almost annually in Cambridge, where, however, they appear to have ceased nesting in 1943. After 1936 black redstarts were more frequently recorded in central London, the first proved breeding being near Westminster Abbey, where a pair reared two broods in 1940 and two more in 1941. On 30 June 1940 I was able to see as many as six black redstarts within half-an-hour in Westminster, one in good song on Farm Street Chapel behind Berkeley Square, and the rest about Westminster Abbey Cloisters. In 1941 and 1942 a pair bred in a bombed building at Wandsworth, and breeding was once more proved at Pett Level by two pairs in each of these years. In 1942 a family party was seen in the City of London on a bombed site in the Cripplegate area, and another in Kensington, while in all about twenty singing males were located. In that year also pairs were found breeding in East Devon (near the Somerset border) and at Whitstable in Kent, and a number of birds were seen in the Medway area (where breeding had occurred in 1940 and 1941) and at Maidstone, St. Leonards-on-Sea, Southampton, Plymouth, Lowestoft, Ely and even as far north as Sheffield. Further expansion was recorded in 1943, when fully seventeen pairs were recorded nesting, three in the City of London, three in Lowestoft, three at Pett Level, two in Southampton, one each in Birmingham, at Bartley near Southampton and near Rochester, and at least one of four pairs present in Dover, and two pairs in Hastings and St. Leonards. Reports of several non-breeding birds came from various parts of London and from Cambridge and Southampton. In 1944, while the number of breeding pairs found was little less than the previous year, there was a heavy reduction in reports of non-breeders, which fell to about one-third of the high level reached in 1942. The City of London had two proved breeding pairs, there were others at Croydon and Stepney, and another resumed nesting on the original site at Wembley just in time to bring off a brood before the ledge was destroyed by a flying bomb. Three pairs again nested at Pett Level, and two more in Suffolk, at Lowestoft and Aldeburgh, but the only others reported outside Greater London were one breeding pair in Southampton, one at Ramsgate, one again at St. Leonards-on-Sea, and a possible breeding record for Coventry.

In 1945 non-breeding records showed a slight rally, but proved breeding pairs fell back to ten, the City of London with two (possibly three) being the only locality from which breeding was reported within the metropolitan area. Breeding pairs were also found at Southampton (two), Ramsgate (two), Dover, Portsmouth and Pett Level (two). There may also have been a nest at Southsea. No locality north of the Thames valley was able to report even a non-breeding black redstart during the 1945 season, except Colchester and Lowestoft.

In 1946 12 or 13 pairs were proved to breed and 13 more pairs or singing males occupied territory without breeding being proved. Five or six pairs bred in the City, three more in Dover, one in Folkestone, one in Ramsgate, one at Pett Level and one at Winchelsea Beach nearby. Thus all the proved breeding, and also all the non-breeding summer residents were either within a single square mile of Central London or along the East Kent and East Sussex coast. A pair seen at Ilkley in May may have been on passage.

In 1947 breeding was again recorded only in the same limited areas, at least 6 pairs nesting in the City, 5 in Dover and 3 at Pett Level, the only other being at Ramsgate, but cocks sang in summer in Birmingham and Bristol.

In 1948 there was both an increase and a spread in the occurrences, between 26 and 29 pairs having been found breeding, and fourteen other singing cocks located. Seven or eight pairs bred in the City and one in Westminster, but Dover took first place with 11 nesting pairs and there were three each at Hastings and Pett Level and one each at Rochester, Gillingham, Brantham in Suffolk, probably Snodland in Kent and possibly Guildford, while cocks sang in Bristol, Birmingham and Cardiff. In 1949, 27 or 28 pairs were reported nesting, distributed between Kent (Folkestone, 5 in Dover, 3 in Margate, Ramsgate and Faversham), Sussex (2 in Hastings and 3 at Pett), the City of London (10 or 11) and West Suffolk (Lakenheath). Non-breeding males sang as far beyond the colonised area as Plymouth, Cambridge, Rhayader, Birmingham and Liverpool. After the breeding season, birds were reported even from Oban and the Farne Islands.

Making all allowances for incomplete information it is clear that the British native black redstarts are still below the threshold of secure survival. The annual influx of settlers from the Continent, small as it must be, is probably still indispensable to the successful colonisation of Britain by the species. Yet the series of cases where particular sites have been used for a number of years and then abandoned suggests that individual colonists and their descendants are tenacious and are very close to producing a sufficient annual surplus to cover their mortality. On the other hand, it is disturbing to find how many of the more successful pioneer efforts have depended on the existence of breeding sites and territories which were not only exceptional in character but also temporary and sheltered by some form of restriction of human access. In such different examples as Wembley Palace of Engineering, Westminster Abbey, Woolwich Arsenal, Cambridge University, and Croydon Airport, we find the common pattern of exceptionally high buildings immediately flanked by suitable small patches of protected open space with some degree of supervision to limit human disturbance. In the blitzed areas of the City, Southampton and the invasion coast towns' wartime conditions temporarily provided something very similar. On coastal cliffs, again, the three factors of suitable inaccessible breeding sites, small patches of suitable foraging territory and shelter from excessive human disturbance recur.

On the Continent the black redstart is a successful and often dominant urban species; why should it be less able to hold its own in Britain? Certainly climate offers no explanation, since numbers of the Continental breeders have long preferred England as winter quarters. One obvious explanation is that in any particular area of the Continent there is a much larger and more accessible reservoir of floating black redstart population to fill any gaps due to misfortune than there can be in this country, where the species is at the extreme tip of its north-westerly range, and the basis for spreading risks has yet to be created. But it is also probably true that conventional British towns contain a larger proportion of excessively built-up areas, lacking in any suitable patches of feeding territory, than those on the Continent, especially in those quarters where the really high buildings so attractive to the black redstart are concentrated.

On the other hand the British suburb and village, with their wide areas of garden and ample cover, probably favour potential competitors such as, perhaps, the robin, enough to render colonisation by black redstarts much more difficult than in Germany or Switzerland. Being by origin a mountain and cliff bird, the black redstart views a town as a series of artificial ravines, which may or may not offer suitable precipices for breeding near enough to suitable foraging patches on which insects and so forth can be picked up in adequate quantity without undue interruption by people, cats, dogs and rival species of birds. The robin on the other hand views the town as a kind of forest heavily cut up and at times hopelessly swamped by sterile and inhospitable buildings, and follows the tapering fingers or isolated splashes of trees and bushes as far inwards towards the town centre as considerations of cover and food supply permit. The black redstart probably has the advantage where there is waste land overlooked by tall buildings and with plenty of insects but few or no bushes and trees, or on windy coasts where trees and bushes may be too exposed to attract possible competitors. Such sites are, however, not very common, and except where some such corporation as a University or Church chooses to maintain a suitable blend of high buildings and open space they are apt to be made uninhabitable, after a life of a few years, by further building. There is, however, some hope for black redstarts in the Town and Country Planning Act 1947, which is likely to encourage redevelopment of the central built-up areas of large towns on lines which have proved favourable for their survival.

The black redstart's spread to Britain is part of an expansion over wide areas of north-western Europe. At Dieppe in the summer of 1920 I found it dwelling both in the chalk cliffs and on buildings such as the Chateau; the cliff birds would go down to forage on the tidal shingle, while the others would search the crannies of masonry, clinging to walls like a wren, and were also expert flycatchers. I saw one hawking for flies from a clothes-line in a garden. The Pett Level birds ate many woodlice, and also butterflies and daddy-long-legs. While black redstarts on the Continent overlap freely with redstarts in gardens and orchards, they rarely range very far from the buildings, walls, or rock faces on which they are based, and I have repeatedly been struck by their fondness for structures which are unusually large or unusually ancient, or both. In an old fortified town such as Pérouges in France, on medieval Rhenish castles and baroque palaces with plenty of statues, even inside the triforium arcade of Altenberg cathedral I have found black redstarts very much at home and I have seen one fly into a French church during the fête-day service. Artificial rocks by the polar bears' enclosure in Cologne zoo, walls of a dismantled German fort, rooftops in town and village are other characteristic habitats. In Alpine regions the black redstart flourishes in the villages along the valley floor, and in July in Dauphiné I have met a pair as high as 9,000 ft. above sea-level on a big moraine, and others above 7,000 ft. The highest I have seen in England were close to the 1,000 ft. contour in north Devon in October.

Black redstarts are strongly territorial in the breeding season and hardly sociable outside it, although I have seen as many as five together in Devon in November and three in the Scillies in March. Not only in the north of their European range but so far south as Switzerland they are summer migrants: in the Rhineland I noted their arrival in the last week of March, immediately after the first wave of chiffchaffs, and they were in full song from then until early July, and again from late August until early October, when they disappeared. In the Belgian Ardennes I have also found them in full song during the last third of September, especially in the early morning and late afternoon. The song period of those breeding in England appears to be very similar, and in their breeding quarters they are summer visitors. I found two in full song on blitzed buildings in the middle of Hull on the afternoon of 25 September 1949: one sang as many as 13 times in a minute. These might have been passage birds. Whether our birds go

PLATE XXXI.

C. W. Teager

House-sparrows harvesting

PLATE XXXII.

C. W. Teager

Starlings

south in winter or join the passage birds staying to winter in south-west England and South Wales, is not yet clear. In either case there is no parallel example of a species breeding in south-east England and wintering in south-west, or being a summer migrant from overseas in the south-east, and at the same time a winter visitor from the Continent in the south-west, with the possible added complication of a few being resident throughout the year.

Quick and nervous in its actions the black redstart likes exposed perches and makes fairly long flights from point to point but in a rather weak, hesitant and pipit-like manner. It also makes quick darts after a fly or moth, hovering quite freely on occasion with the chestnut tail turned down to serve as a brake. One of the Westminster Abbey birds watched on 30 June 1940 puzzled me by suddenly without warning lying down on the grass of the cloister garth like a dead bird, with neck and body stretched along the ground, tail horizontal, wings slightly open and body plumage fluffed out and slightly heaving. He lay like this for about two minutes, then suddenly jumped up and hopped off as alert as ever, making a slight noise like the scratchy note of the song. There was no apparent external cause, and no sign of fright.

The black redstart's song is characteristic and remarkably penetrating, and is often the best or even the only way of detecting the bird's presence in an area unfavourable for sight observation, such as narrow streets flanked by high buildings. In such areas any song which does not clearly come from a cagebird or a starling should be most carefully investigated. While the black redstart's song varies in detail it is always penetrating, warbling and brief, lasting not more than about 4 seconds. The warbling notes are short, brisk and usually unmusical, and are often followed by an extraordinary scratchy or pebbly note which has been described as like a handful of small metal balls shaken together. When complete the song ends with a more musical brief warble. I have timed 5 complete utterances or up to 12 or 13 incomplete phrases in a minute. I have heard the song given on the wing, and also during heavy rain. In Germany at close quarters I have often heard a stuttering rattle preceding the song. The note of alarm or excitement is normally a triple "tit-tic-tac" or "tchee tuk-tuk", the pause between the first and second notes being about double that between the second and third. Winter birds are usually silent.

In Germany I have found a half-built nest under the eaves of a Rhenish castle on 15 April, within three weeks of the first arrivals, but in England a number of birds do not even take up breeding territory until late April or May, leaving themselves little time for rearing the two broods which are normal. In Germany I have seen a nest on top of a coroneted stone sculpture ornamenting a door-porch about eleven feet up in a narrow crowded street with trams passing continually, and the Birmingham nest was above the main entrance to the University, but in Britain the preference seems usually to be for more secluded sites. Some nests are in narrow holes in rocks or masonry, others on open ledges or rafters, others within sheds or other accessible buildings. A pair at Lowestoft nested in an old fishtrunk nailed to a post inside an open-fronted store, while a pair at St. Leonards used a hole quite 6 ft. below ground level in the remains of a bombed cellar. Except on cliffs black redstarts appear to be almost entirely dependent on human action for the creation of their nesting sites. The nest, built by the hen, is loose and untidy, made of grass fibre, and moss, lined with hair and feathers. The eggs are white like those of most hole-nesters, and there are four to six in a clutch. They are incubated by the hen in about thirteen days and the brood is normally fledged in sixteen to eighteen days more. In England small broods of three or four young seem most frequent.

In size and build the black redstart, like the common redstart, is like a slenderer robin with a chestnut or bright rusty tail and under-tail-coverts, the rest of the plumage both above and below being ashy or brownish-grey in most birds seen in England, which are females or first-year males. The fully adult male is a much more handsome bird with jet-black throat and breast and a broad white wing-patch. Both in Germany in June and in Devon in November I have seen an old cock with a hoary white semi-lunar mark on the forehead, like a common redstart. While the young common redstart begins life like the robin, well spotted above and below as a reminder of its thrush relations, the black redstart leaves the nest so leaden in plumage (except for the tail) that the black spots on it are quite inconspicuous. The bill at this stage is yellow; later it becomes black, like the legs and feet.

SWALLOW

(Pl. 36, 37, pps. 173, 208)

Although almost entirely invisible, since it consists of large tracts of air surrounding the nest or roost, the swallow's habitat is fairly well defined. It is, in summer quarters, a rather shallow layer of air from immediately over the land or water up to normally only some five hundred feet or so above the surface, especially over marshes, rivers and inland waters, farms, sewage farms, villages, gardens, parkland and the slopes of hills. It is mainly a lowland species, but will go up foraging over suitable moors to a fair altitude; I have seen it on moors in Ireland up to 1,300 ft. At high elevations the swift often seems to take its place. Coming up the Mont Cenis line in Italy in July 1927 I made a sectional sample count over three sections; on the first, of some 13 miles from near Turin in the Piedmont plain up to about 1,250 ft. I checked 34 swallows and no swifts; on the second of some 16 miles of upland valley rising to about 1,800 ft. the score was 16 swallows and 12 swifts, while on the final 18 miles of open track rising to 4,300 ft. no more swallows were seen but 122 swifts. In Tirol I have found swallows common up to about 3,000 ft. in the valleys. It is unusual to see swallows spending much time over considerable blocks of woodland or towns of any great size.

Sample censuses in Britain have shown that in the most favourable areas swallows slightly exceed 40 breeding pairs per 1,000 acres, showing about the same maximum density as house-martins. The swallow, is however, more numerous, owing to its wider distribution over the country as a whole. Even where food and nest-sites are most plentiful swallows never crowd their homes together closely as house-martins, although they make no effort to hold separate feeding territories. Although quite sociable they often forage and migrate alone.

Spending so much time on the wing and being so skilful at evasive action the swallow is little exposed to enemies except at the nesting-place, where it is liable to be killed by a cat, a rat or an owl. Cold weather sometimes inflicts heavy losses on swallows, although they can stand a good deal provided their food is not cut off.

Swallows suffer greatly from parasites infesting their bodies and their nests. The red mite *Dermanyssus gallinae* has been found in countless numbers in English and Welsh nests, in some cases being suspected of causing death of the young. A blue-bottle fly *Protocalliphora caerulea* is parasitic as a larva on nestling swallows, to which it attaches itself by small hooks, sucking blood and apparently causing death in some cases. These bluebottles are themselves victimised by a brilliant green Chalcid *Mormoniella vitripennis*, while the red mites are eaten by other insects, and various moths even feed on feathers used as nestlining.

It is curious, in view of the swallow's migrations, that French birds appear to have several species of common parasites different from those found on British birds. This points to effective segregation of the British native stock.

Unlike house-martins, swallows are rarely evicted from their nests by house-sparrows, which however often take over and use old ones. Wrens do the same. The Report on the Swallow Enquiry records how two old swallows' nests at Tabley, Cheshire, were occupied one by a wren, the other by robins, which had hatched four young when the swallows returned. Then the robins were driven away and each of the young was found dead with a wound over the eye like a peck, the cause not being proved. All the nests were then removed, and the robins again built on a board fixed under the swallows' nest until after they had fed four young for eight or nine days the swallows again made them desert. Fortunately such feuds seem to be rare. One unlucky robin in Kent took over a swallow's nest and then had to rear a young cuckoo in it.

Although the swallow has for generations been famous as a long-distance migrant, the argument whether it really travelled south in winter or hibernated in chalkpits and other lurking-places, remained open until after Gilbert White's time, and it is only within living memory that reporting the finding of torpid swallows in winter has gone completely out of fashion.

Bird-marking has told us much about the swallow's journeys, but there is still much more to be discovered. All but one of the British-ringed swallows so far recovered in winter quarters have been found in the areas of South Africa fringing the Indian Ocean. They go in fact from the remotest corner of Europe to the remotest corner of Africa as a habit. Some of the Danish birds and two from elsewhere in Europe have been found in this same area, but all except one of the German winter swallows recovered have been close to the Equator in central or western Africa. The indications are that German swallows reach Africa on a fairly broad front between Malta and the Cape Verde islands, and that our own birds travel through France and Spain. As the journey takes some weeks quite a large proportion of a swallow's life is spent in travelling.

The earliest to arrive usually reach England before the end of March. The great majority come in April, but some of our swallows go on straggling in through May, when passage birds are also frequent.

PLATE 37

Eric Hosking

Swallow in flight

Eric Hosking

a. House-martin collecting mud

T. M. Fowler

b. Sand-martin entering nest tunnel

Hardly have these finished passing than the first southward movements of our breeding birds begin in July, but it is in August and even early September that the number of swallows on the wing in Britain reaches its annual peak as fledgelings and passage migrants swamp the nucleus element of breeding adults. The big exodus goes on through September and much of October, and swallows in November, although uncommon, are less infrequent than almost any other summer migrants. I have never myself seen one in England between 4 November (Devon) and 23 March (Somerset) but winter swallows have been reliably reported, and an occasional lucky bird may survive here to see the rest return in spring.

Swallows of our own race reach their north-westernmost regular breeding areas within the British Isles, spreading thence across practically the entire depth and breadth of continental Europe and overlapping into North Africa and south-western and western Asia. Other closely allied migratory races breed throughout the rest of Asia (except south of the Himalayas and of Indo-China) and in North America, where the birds have a deep buff breast and are called barn-swallows. Egypt alone has a race which does not migrate.

Although swallows look so perfectly at home and so adept at flying they are less aerial than swifts. During the breeding season most of their perching is done on or just by the nest, and it is on first arrival or after the young are reared that they are most often seen perching in the open, especially on twiggy, not too high trees such as an apple, a walnut, or a late-leafing ash. A swallow singing on a low ash tree near Birmingham in June 1947 was sharing the tree with a cock willow-warbler and yellowhammer, also singing, another pair of yellow-hammers and a pair of pied woodpeckers. In Cornwall in September I have seen a dozen swallows perching, accompanied by a house-martin and sand-martin and chaffinches and yellowhammers, on sprays of wild rose, hawthorn, honeysuckle, hazel and thistleheads along a wild hedge; others settled on a rick of straw. On the whole, however, swallows seem to prefer perching on overhead wires or on buildings, and in autumn they will do this in large numbers. I have noticed them settling on the ground in early June in Suffolk, apparently collecting flies, and once on Midsummer Day on the grassy crest of Midsummer Hill in the Malverns; I have also seen low-flying birds settle on the close-mown green of a golf links in Essex in September, one of them squatting like a sitting bird. While they cannot actually settle on the water swallows like to sip it as they skim. At Corofin in County Clare I have seen them warbling with pleasure as they plunged towards a clear pool and dipped themselves in the water flying, so that they made a great splash.

In flight the swallow displays a mastery and a sense of style matched by few of the world's birds. Each wing-stroke is a pleasure to watch. Yet few species average a slower airspeed, and even in peril or in the chase the swallow's acceleration is by no means outstanding. The long pointed wings, admirably adapted for sudden turns or for flying through a slit a couple of inches wide, are less suited to soaring or gliding; hovering is done with great skill but not very often, except when feeding young on the wing. The food consists entirely of small flying insects.

The voice of the swallow is twittering, neither very musical nor very shrill or harsh, although under excitement or when alarmed it rises to a high-pitched and fairly loud shrill cry. The song is uttered freely on the wing, on a song-stand or even inside a boathouse, but in volume and development it hardly ranks as more than a subsong, being delivered mainly with half-closed bill so that it rarely carries beyond forty yards, and being composed of a stream of slight variants on the ordinary notes, briskly and engagingly delivered.

The nest, built of mud reinforced by bits of straw and lined with feathers, is fairly shallow and open at the top, and therefore cannot survive without some effective weather protection overhead, usually provided by a building. In Tirol I have seen many nests, of the normal saucer-shaped type, under eaves in typical house-martin situations, but in England the vast majority of nests are on rafters or ledges inside buildings. Farm outhouses are strongly favoured, but swallows will also use porches of churches or even dwelling-houses, and at one ancient house near Oxford where windows or doors were open throughout the summer, swallows nested in the scullery, and regularly flew through the large hall which was used as a sitting-room. The British Trust for Ornithology's Swallow Enquiry brought out the interesting fact that clutches of eggs and broods of young increase in size towards the north of Britain, especially in June and July when daylight is longest. The normal clutch is five, but in the north six are not infrequent. Over the whole country and the whole season broods averaged four young, but round Ullswater in the Lake District over 40 per cent of broods recorded contained as many as 6 young.

The swallow is perhaps the most distinctive of all passerine birds. The strikingly elongated and narrowed wings and outer tail-feathers, the shortened and weakened legs, and the widened and flattened beak are all clearly modified to serve the swallow's function in life of catching with the greatest of ease enormous quantities of flies. The male swallow, although in body one of our smallest and lightest birds, has a wing as long as a blackbird's and a tail even longer, but only half as long a bill and one-third as long a tarsus. To this slender and graceful build the swallow adds a handsome plumage, glossy blue-black above (without any of the brown tinge visible on the house-martin) and white to pinkish-buff below, with chestnut forehead and throat, a fairly broad blue band below the latter and white "windows" near the tips of the tail-feathers, of which the outermost are prolonged, especially in the male, into stiff narrow streamers which produce the famous forked swallowtail outline. The flat triangular bill and the legs and feet are black, the iris dark brown. The female is duller and shorter-tailed than the male and the young considerably more so, but there is no essential difference in plumage.

HOUSE-MARTIN

(Pl. 11a, 38a, pps. 17, 209)

Originally a cliff-breeder, and presumably therefore tied to coasts and inland escarpments, the house-martin has greatly enlarged its distribution by adapting itself to nesting upon buildings. Sometimes it picks a site under the roofs, but generally on an outside wall protected by some overhanging ledge, most often under the eaves of houses. But, unlike the swallow and swift, in taking to buildings it has by no means abandoned its more primitive haunts. On the north Pembrokeshire coast near Newport all the local birds were in 1945 concentrated in two small breeding colonies on seacliffs, one of about twenty and the other of about three nests, the only other breeding martins in this immediate area being based well inland, ignoring the houses near the shore. The number breeding on cliffs in the British Isles must still be substantial. In the French Alps in July I have seen one hawking flies over the moraine of a glacier at about 9,000 ft.; it kept visiting some crags where owing to the glare of the sunlit snow I could not follow it, but I feel sure it had a nest there.

Perhaps because of its origins the martin has retained a marked liking for water and a marked tendency to nest and live in colonies

Often spending nearly all the daylight hours on the wing, where it habitually takes its insect prey higher above the ground than any other bird except the swift, the martin seems to have a much wider choice of possible homes than it chooses to exercise. It has a curiously patchy and erratic distribution, being common in some places and quite scarce in others which superficially look no less suitable. It clings to some ancient breeding places and deserts others, showing no hesitation in colonising new buildings with little apparent attraction, including even certain houses of design identical with others which it ignores. It does not reach the Outer Hebrides, nor usually Shetland, and is somewhat southerly in its distribution. Sample censuses summarised in the Report of the Swallow Enquiry 1934 and 1935 show densities in various parts of England and Wales varying from nil to 40 pairs of house-martins per thousand acres. In view of these abrupt and inexplicable variations it is more than usually difficult to estimate how common the house-martin is, but it may quite possibly rank in summer among our two dozen most plentiful species.

While house-martins not uncommonly nest singly they prefer to form colonies of up to forty or fifty pairs, and in very favourable conditions even more. There is no question of their holding breeding territory, as the air is their hunting-ground, and it is quite common for nests to be touching one another, even when there is plenty of room unoccupied. Martins have been proved by ringing to enter other martins' nests, and they forage, collect mud, sun themselves and travel together. They do not seem to be jealous or quarrelsome birds like many colonial nesters, and although they do not usually nest on the same building as swallows this is probably due to the fact that they like to breed on the outside of buildings while swallows like to be under the roof, and therefore usually go for outbuildings rather than dwellings. The one feud in which house-martins are chronically engaged is with house-sparrows which invade their nests, and are recorded as making holes in them, and even ejecting eggs and young. On some occasions the house-martins seem to put up little or no resistance, but on others they defend themselves successfully.

If martins are themselves mildly parasitic on man for breeding sites they in turn suffer from all kinds of parasitism. A single house-martin's nest collected near Weybridge by P. A. D. Hollom, was found to contain 452 living parasites, while two other nests contained 99 and 182 of them. These were fleas, bird-flies and martin-bugs, and fowl-mites—all parasites described as feeding "at very frequent intervals only on the blood of their host, which must suffer in consequence a considerable loss of blood". But in addition to these nest parasites the martin suffers from at least three species of lice and four species of mites which are entirely body parasites. Watching the graceful aerial movements of the house-martins no one would guess what a heavy burden of bloodsuckers they have to bear.

The house-martin is a long-distance migrant, represented by various subspecies across a broad but not very deep breeding zone stretching from County Kerry eastwards across the Old World to Japan. In England it is one of the later spring arrivals, the main body often not settling in until May, although mid-April vanguard birds are frequent and March arrivals are not entirely unknown. Unlike most late arrivals they are also late in going south; although emigration begins in August considerable numbers remain until October, and November stragglers are not exceptional, even December occurrences being recorded. The martin has, however, never been shown to survive an English winter. Although evidence from ringing so far gives little help, our breeding martins are probably among those which winter in South Africa.

House-martins are rarely seen on the ground except when collecting mud for nests, and then they usually alight by the water's edge and take off again from the same spot with little or no attempt to walk. On 21 July 1940 on a very steep and much frequented patch of Box Hill, Surrey, I counted up to 17 martins at a time dotted in several groups over a few yards of the short dry chalky turf, apparently resting and sunning themselves. Several at least were young. This was at 6.30 p.m. Greenwich time. As in the Newmarket case described on page 10 this happened in a particularly warm summer and the slope chosen got the full benefit of the sun.

In Suffolk in June 1948 I found house-martins in one small area apparently feeding on the ground, on the fine tilth of a newly harrowed and manured field, and the next day again on a similar field adjoining. This very local experiment was being tried by swallows and sand-martins as well as house-martins, but house-martins were in the majority and I suspect the practice had originated through house-martins which constantly visited a nearby ford to pick up mud for their nests. These birds may well have found that juicy insects could readily be picked up on the ground and so been emboldened to spread the search to neighbouring fields on which they alighted only for short periods with a great show of nervousness and usually with one or two companions of their species to give moral support. Parties and flocks often settle on overhead wires during the late summer and autumn, and I have sometimes though rarely seen them perch on trees at this season. Although so little given to perching they are expert at such difficult feats as flying up and clinging to a vertical wall.

The flight is easy, almost leisurely and often gliding, without the violent energy of the swift or the finished style of the swallow. House-martins are capable of ranging far afield, but usually choose to remain within sight and hearing of the breeding-place, gathering flies and other insects in endless cruising circles, sometimes at heights of several hundred feet above the ground. The voice, much used, is a sure guide once learnt; it is an easy thick chirrup uttered more in the tone of a remark than an exclamation. There is a shrill rather high-pitched alarm, and once at Thursley, Surrey, I several times heard a much softer variant of this—a piping, almost bullfinch-like note, uttered by a bird flying up to a nest and sheering off at the last moment. The song, or rather subsong, is a pleasant easy inward warbling, based on the call-notes, uttered on the wing, or on a perch or at the nest at any time during the bird's stay with us.

House-martins, both adults and juveniles, roost in the nest, and it is astonishing how many can contrive to crowd in to the small space, especially in view of the irritation which they must suffer from parasites. In Switzerland up to a dozen have frequently been found in one nest.

As soon as they reach summer quarters the martins start repairing last year's nests (if these are in good enough order and not seized by sparrows) or otherwise building new ones. In shape the nest is like no other found in Britain, being a deep coconut-like structure of mud, usually with an open top closely sheltered by eaves or some other shallow overhang of rock or masonry, or occasionally thatch, and with a small entrance hole. Sometimes when an unusual site is adopted the nest may be a regular ovoid like a wasp's nest with a circular hole at the narrow top. I have seen one like this beneath a high stone railway viaduct near Cork. A rather similar one in the ruins of Rievaulx Abbey, Yorkshire, was shaped like a long-tailed tit's nest with rounded dome and a hole at the top of one side. In a village near Cheltenham a pair built on top of the half-open casement window of an inn; it began as a cup like a swallow's, but when completed was supported above by unusually broad eaves, having no point of attachment at the sides; in the end it was destroyed by the window being moved. Where groups of nests exist it is not unusual to find that less than half of them are actually occupied by martins while some have been taken over by sparrows and the rest are empty. The eggs, like most laid in holes, are white, four or five being the usual clutch; two broods are normal and three are not uncommon. Both sexes incubate; eggs hatch in 14–15 days and the young fly about three weeks later. In September 1924 in Hertfordshire, watching the young being fed in a solitary nest I found that several—apparently five—different birds were bringing food, and from their browner backs some were young of the year, probably the first brood.

Generically house-martins are separated from swallows most conspicuously by having their legs and feet feathered to the toes. They look slender, although fuller-chested and more blunt-tailed than swallows; in flight the tail has always seemed to me to have an almost fish-like outline, and the wings seem disproportionately deep from the front back to the rear edge, as well as being somewhat blunt-tipped. The plumage is simple, the underparts of the body and the broad rump bar being pure white at all ages, while the upperparts from the forehead to the tips of the wings and tail are blue-black shading to dark brown, juveniles being readily picked out by being so much browner. The length is about 5 inches, the bill, tail and wings all being shorter than the swallow's, but the legs averaging a shade longer.

SWIFT

(Pl. 39, p. 224)

No other British bird is so aerial as the swift, a species which not many people can claim ever to have seen at rest. Except at the nest the swift's life is spent flying in a layer of air which stretches from just above the ground to fully 1,000 ft. above it. This layer is rather deeper than a swallow's, but is more briefly tenanted because to support the types of insect life on which the swift depends it must not normally drop to a temperature much below 50 deg. Fahrenheit. Enormous expanses of April and September air which provide good pasturage for swallows in northern Europe are apparently not rich enough to sustain swifts.

Even in those dozen to sixteen weeks which are all that most of our British swifts spend each year in their native land they would find it hard to withstand the climate but for their remarkable capacity to concentrate quickly in the most sheltered and favourable spots, sometimes many miles distant. This capacity is their salvation when the rigours of a British summer make high-flying insects suddenly disappear over wide areas. Without accepting the assertion that swifts sometimes spend whole summer nights high in the air there is little doubt that they spend a greater proportion of the twenty-four hours on the wing than any other British bird. Early up and late to roost, with a high cruising speed, a swift at midsummer probably covers over 500 and perhaps even up to 800 miles in the day's work. It is a sobering thought that no human machine yet designed on land or water and very few aircraft can be operated reliably, week in and week out, over anything approaching the mileages which swifts are accustomed to cover, not only when they are migrating but also when they are just pottering about.

It is difficult for us with our sedentary outlooks to picture the daily routine of so mobile a creature. The astonishing numbers of swifts which turn up in the middle of the breeding season far from their nearest possible nesting places before or after a storm or on a sudden drop in temperature give us some clue, but we have as yet no knowledge how far these birds move about. When we see large numbers suddenly over St. James's Park lake, or over some remote high moorland we know they have come miles, but cannot guess how many.

Partly on account of superior mobility, and probably also because they feed on higher-flying insects less affected by the nature of the ground underneath, swifts are frequently seen foraging over areas which swallows and martins generally avoid, such as London and other large towns, high fells and moors, and large woodlands. On the other hand they seem to fall back on foraging low over water mainly as a last resort in poor weather, whereas swallows are fond of it in all weathers.

Like the petrels and shearwaters, and unlike every other British bird, the swift never willingly comes to rest on land except in, or entering or leaving a hole. While the hole selected is now nearly always in a man-made site, almost any dark enough and inaccessible hole will serve for the rough crude nest made of any kind of airborne litter collected by the bird in flight and compounded together with its saliva.

On migration and at play swifts are sociable, and they concentrate their nests in particular buildings probably oftener than can be accounted for by lack of somewhere else to go. But they are hardly to be ranked among the keenly gregarious species, and solitary or widely dispersed swifts are just as frequently seen as swifts in parties or flocks.

Owing to their conspicuous ways and extreme mobility swifts probably tend to appear more abundant than they really are, although in some places their numbers are certainly large (but never as large in England as in many towns on the Continent).

Apart from the British climate the swift has no serious enemies in Britain. While swallows need to rear two or three broods in a season, perhaps totalling eight to ten young, swifts manage to maintain their numbers equally well on a single brood of no more than three, and often only two. No other small land bird breeding here can show a comparable record. Its life, which looks to us so strenuous and precarious, is shown by mortality rates to be in practice relatively secure. Even in getting food the swift has no regular competitors, since its foraging area, except in bad weather, is usually above or beyond that of any other bird living on insects caught flying.

While other summer birds such as the swallow and cuckoo are more famous as migrants it is the swift whose migrations are most open and readily observed. This point was exploited and confirmed in the all-day checks of swift migration carried out in May 1947 and 1948 by British Trust for Ornithology observers under the direction of H. G. Hurrell. In 1947 more than 500 observers reported on the movements of about 5,000 swifts, three-quarters of which were heading north. Watching on that occasion from a point far out in the mouth of the Thames with the nearest coast of Kent 13 miles to the south and the nearest coasts of Essex 10 miles west and 15 miles north, I saw none from 3.30 a.m. Greenwich time when watch started in the chilly grey dawn until 4.40 when a swift flew past at 50 feet heading west for the nearest Essex coast, by Foulness. At 5.30 another passed W.S.W. towards Shoeburyness at 100 ft. followed almost at once by the first swallow and a probable house-martin. Two more swifts passed W.N.W. half-an-hour later, and only two more were seen, at intervals of a couple of hours, although swallows and house-martins were much more frequent. In 1948 the general direction of flight noted by observers proved to be somewhat more easterly, probably owing to a tendency to head into the wind. In 1949 a more extended watch showed the main arrival in England in the second week in May, and in France I saw numbers just arriving from across the Mediterranean as late as 10th May. Few swifts reach Britain before the second half of April, and the earliest I have ever myself seen crossed the Thames at Long Ditton on 20 April 1925. On the Rhine, however, in the same latitude as Sussex, I have seen a single bird as early as 12 April and a flock of nearly 40 on the 17th. By mid-July the exodus is already beginning, and at the end of the month when flocks roll up to large numbers I have seen under Bredon Hill in Worcestershire a migratory band of possibly as many as six hundred birds. Often, however, the actual migratory movement consists of single birds and small parties travelling on a broad front; once in August on a journey across South Wales I noticed southbound swifts all the way between the Severn and Fishguard. The latest I have ever seen in England was on 3 October in Suffolk, and occasional stragglers have been recorded later in October, and exceptionally even until December.

Swifts, taking the typical and the eastern races together, breed from County Kerry and the Highlands eastward across Europe and Asia to China, extending nearly as far north as the swallow but with a more southerly centre of gravity; they breed nearly all round the Mediterranean, and winter in southern Africa. Owing to their habits, however, it has not yet been possible to trace their migrations by bird-marking.

In flight the long stiff narrow wings give the appearance of being unflexed and inflexible in every direction: masterly as the flight is it does not look graceful. Regarded aeronautically the aspect ratio of the wings is remarkably high, and the wing-loading is considerable for so small a bird, while the elimination of virtually all perching activities except sitting on the nest or at the roost has enabled stream-lining to be carried to a high degree of perfection and drag to be correspondingly cut. In style of flight the swift perhaps comes nearer to the fulmar than to any other British bird, but is much faster and less given to gliding, apparently commanding a much higher amount of power in relation to weight. Swifts have been timed at 70 and even 100 miles per hour, but slower speeds are normally adopted, although few birds are less capable of really slow flight and a hovering swift is unthinkable. Not only the acceleration but the rate of climb and the turning ability are outstanding, and it is suspected, although not proved that swifts have the power of moving one wing faster than the other in swerving. We have no bird which takes such evident and continual delight in headlong chases, even round blind corners, and cases are on record where swifts so engaged have collided and killed one another.

Swifts have a passion for mating in flight, the male riding with sharply raised motionless wings on the back of the female who holds hers horizontally, while one or both birds utter a curious quickened ecstatic scream quite distinct from the normal cry and serving to alert the experienced observer to what is going on overhead. A steep dive often follows, either before or after breaking away. I have seen this in Hampshire as early as 2 May, within ten days of first noting the arrival, but it seems to be most frequent some three weeks later.

It is a favourite speculation among some ornithologists whether swifts can sleep, or at least spend the night on the wing. Undoubtedly they sometimes ascend to great heights about sunset; once in mid-June I heard swifts screaming overhead after sunset too high to be visible. But from the nature of things proof of nights spent in the air is elusive, and it seems an improbable habit, even for so strange a bird.

The basic note of the swift is a high-pitched, more or less prolonged scream, of great penetrating and carrying power, and unmistakeable once learnt. Although usually uttered on the wing it may be heard from nests too. In the chase a more excited and less harsh squeal is freely used, and the pairing note already mentioned is also quite distinct. A persistent "chick-chick-chick-chick-chick" at the nest is also reported.

The two or three dull white eggs are laid late, often not until June, and take eighteen or nineteen days to hatch, after which the young remain in the nest for five, six or even eight weeks, with a journey to Africa facing them almost immediately they are fledged.

The swift is the only British representative of the order which includes the humming-birds. It is very different from most other birds in most respects, from its four strong forward-pointing toes to its exceptionally thick skin. Although shorter in total length than a swallow it has wings each a full two inches longer, and in fact as long as the bird itself. The bill and tarsus are just about the same length as a swallow's, despite the prevalent idea that the swift suffers from having only rudimentary legs.

In the field the swift is a plain sooty brown bird with a whitish throat and a black bill. The need for this sombre colouring is presumably to give protection in the dim nesting and roosting places where it is most vulnerable, just as the nightjar's plumage is adapted to its nesting environment.

CHAPTER 9

BIRDS AMONG PEOPLE

Great Britain is an island of some fifty-six million acres in which nearly fifty million people are living among roughly twice as many wild birds. In this book some attempt has been made to explore or to illustrate the variety of the problems which arise out of this mingling of people among birds and of birds among people. These problems are largely new. They are rapidly changing and are multiplying every decade. More people are making more changes of more different kinds to more and more acres than ever before. More birds, and more species of birds, are being affected in more ways and at a faster pace than ever before. Most of this impact is unintended and indirect, much of it unrecognized and unobserved. Yet by a surprising coincidence this exceptional impact of people among birds in mid-twentieth century Britain coincides with an exceptional growth of interest in bird life. In few if any other large countries and at no other period could the present scale and variety of human impact on bird life in Britain be matched, and in no other country is there, or has there been, so widespread an informed interest in birds. It is the purpose of this final chapter to review the opportunities and responsibilities which flow from these facts.

Outstanding among the opportunities are those for enjoyment and those for research. Outstanding among the responsibilities are those for research and those for protection. If such vast and sudden changes had developed in a nation mainly uninterested in bird life, as many nations are, a few ornithologists might have had glimpses of what was going on, but could scarcely have hoped to chronicle it, let alone to influence it. On the other hand if an equally lively and widespread enthusiasm for birds had sprung up in some country not undergoing any great changes the opportunities for research and for protection would have been far less, although it is easily arguable that the opportunities for enjoyment would have been greater. As it is, nothing in British ornithology is more urgent than that we should obtain and use, greater knowledge of the problems of birds among people. The greatest difficulty in preventing or mitigating harm to bird life is not inability to secure action but ignorance of the facts on which action should be based. More and better bird-watching is the way to more and better bird-protection. The better we understand how birds live, the better we can ensure that they will not be deprived of the opportunity for living.

As earlier chapters have shown, it is wrong to assume that the march of civilisation is in itself harmful to bird life. Many of the changes which it brings are mainly or wholly beneficial. Where the builder has left room for the gardener to follow building has probably usually increased rather than diminished the density of wild bird population. In the short run there has often been some impoverishment in the number of species, but it would be rash to conclude that built-up areas with enough mature gardens cannot compare favourably in character and variety of bird life with many of the habitats from which the land was taken over. Here is a subject on which more observation and study is required. We know something of the ecological succession which occurs when heathland is afforested and the growing trees attract and in time repel successive waves of bird colonists until the climax of tall forest is reached. We know very little, however, of the ecological succession from "ripe building land" to raw new construction and thence to treeless gardens, gardens with shrubs and saplings, and finally roads and gardens with mature trees, creepers and decaying timber. Creation of new parks, parkways and amenity planting is undoubtedly often beneficial to birds, but we have very few and imperfect records of the process of colonisation and only limited knowledge of the types of planting and of landscape gardening which are most favourable to birds.

The favourable effects upon bird life of the bombardment of our cities from the air have been much better observed and documented than the effects of builders and contractors, and we know which species are attracted (notably black redstarts in summer and grey wagtails in autumn) and the conditions which attract them. We are also becoming reasonably well informed of the changes brought about by new reservoirs, which are greatly increasing the frequency of visits by many hitherto rare birds in many parts of the country and are to some extent offsetting the regrettable failure of sewage engineers to continue operating sewage farms on the early twentieth-century model, which was such an inspired if unintentional contribution to the art of attracting rare birds.

We also know something (though still far too little) of the effects of afforestation, which vary enormously according to the previous bird population of the site, the variety of different species and ages of trees planted over a period, and the size and layout of the plantations. Except where the site selected for afforestation is a habitat of uncommon and interesting species which are bound to be displaced, or is likely to create an unmanageable reservoir of crows near some vital nature reserve, or involves replacing interesting broad-leaved woodland by conifer plantations there is in general little reason to fear that afforestation will have adverse effects on the density, variety or interest of our birds, and in many cases it is likely to prove a definite advantage.

The great expansion of shipping and fisheries during the past hundred years is the only other example which need be given here of a human activity which has had immense unintended and until very recently unsuspected effects on the bird life not only of the seas, but in the case of several species of gulls, of the landward areas also. Coast-based gulls, thus encouraged to multiply, have invaded farms far inland and have increasingly colonised even our towns in most months of the year.

If earlier human generations have reduced the numbers and variety of woodland and of water birds by clearance and drainage the present generation is well on the way to making good the losses of these broad habitat types. The variety of suitable haunts for both woodland and water birds in England has probably never been greater and the area, although much diminished over the centuries, is now rapidly increasing in both cases. In the nineteenth-century urban and industrial development happened to take a form highly unfavourable to bird life, and this gave rise to exaggerated ideas about the inherent incompatibility of civilisation with a rich avifauna, and about the necessity for rigidly segregated "bird sanctuaries" to preserve some remnant of bird life. The absurdity of this view is shown by the fact that the Metropolitan Police District is rapidly becoming one of the best mainland areas in Great Britain for seeing rare birds, several of which, like the black redstart, little ringed plover and smew can be observed more readily here than almost anywhere in the country.

With minor exceptions (such as railway marshalling yards, some types of factory and depot and parts of large airfields) civilisation is not now creating in the British Isles further areas inherently adverse to bird life. The difficulties we are making for birds are of other kinds. One of the worst is the continued encroachment on marshes, lowland heaths and other types of habitat which are now so restricted that further losses of any not already thoroughly spoilt are liable to be serious. We can and do present the woodland and water birds with acceptable substitutes for felled primitive forest and drained meres. but we do little or nothing to mitigate the loss to marsh birds of the drained marshes or the loss to heath birds of built-over, ploughed up, bulldozed or afforested heath and common. Flooding for wartime purposes and neglect of some drains and coast defences has made slight additions recently to our remnants of suitable marshes, of which the bittern, bearded tit and avocet have been quick to take advantage, but agricultural and other authorities tend to look upon marshes as a challenge and reproach, to be tackled with draglines and bulldozers at the earliest opportunity. Until agreement has been reached on the permanent preservation of an adequate fragment of our remaining marshlands on scientific and amenity grounds the future of our marsh birds must be one of the main anxieties of British bird-watchers.

Heath and common lands are often better protected by vested interests and by amenity usage, but being technically waste such areas are naturally the first to be picked by officers looking for somewhere to let off ammunition and explosives, drop bombs, practice tanks and infantry or create some alternative form of hell on earth. While the military mind is busy with new ways of leaving our heaths trampled, scarred, burnt and pitted, highway surveyors dream of the happy time when they will be allowed to get ahead with those blue-prints for driving twin-track arterial roads across them, and in yet other offices foresters persuade themselves how much more satisfactory the entire area would look under a thick stand of corsican pine and sitka spruce, with the necessary straight firebreaks at intervals. Meanwhile the Dartford Warbler, stone-curlew, wheatear, woodlark and other heath birds eke out a precarious existence, threatened by excessive disturbance and persecution in some areas while in others neglect and removal of grazing animals allows the heaths to become too overgrown for their needs.

A third group of habitats which urgently need protecting from further disturbance and alienation is the small series of sand-dunes, isolated beaches and spits, lagoons and saltings, bird-cliffs and islets dotted here and there round our coasts, which are important as breeding sites or staging posts for our sea and shore birds. In some cases the freedom from disturbance of a small patch of beach for a few weeks in the year may be the condition of survival for an important breeding colony or for the continued use of a bay or estuary by flocks halting on migration.

The total area of these remnants of marsh, heath and coast whose preservation is vital for these purposes is relatively small—probably under one-quarter of one per cent of the land surface of Great Britain. What makes them particularly difficult to safeguard is that they are often places which most non-ornithologists are accustomed to regard as useless and calling for, or at least available for, some type of exploitation, while unlike built-over fields and felled woodlands they cannot readily be replaced because their use by birds is so specialised, and their environment irreplaceable in any period worth considering. The survey and observation of all these sites in order that we may be fully informed about their birds and other fauna and flora and the essential conditions of their preservation is so urgent that every year lost in completing it means further irreparable damage to some of our most interesting bird life.

Encroachment on hill and mountain land is not yet serious, although it may well become so with the development of hill farming and of new sources of fuel and power. It is not improbable that by the end of this century some of the main peat bogs will have been intensively exploited for conversion into electric power by gas turbines, and that a large number of hills will have become crowned by tall metal windmills, while many deep new artificial lakes for hydro-electricity will certainly have been developed in mountain areas, and the cattle and sheep populations of the higher slopes may be greatly increased. Such encroachments will no doubt benefit a number of birds, but they will tend to increase the human population and to make for more disturbance of some of the areas at present most lonely and free from human infestation, thus adding to the difficulties of those birds most intolerant of approach.

John Markham

Swift clinging to wall. Taken by flashlight inside a Norfolk house

Eric Hosking

Woodlark with young on a Suffolk heath

In addition to safeguarding scarce habitats against alienation and encroachment it is important for our birds, as for ourselves, that we should use land without unnecessary destruction and with the aim of creating new amenities in place of those which we have to destroy. Polluted air and polluted waters, tightly-packed buildings with few or no spaces, indiscriminate felling of trees and grubbing-up of hedges, and the conversion of mile upon mile of coast into asphalted and terraced promenades are undoubtedly bad for bird life, but it is questionable whether they are any worse for birds than they are for ourselves, and in removing faults which make this country uncongenial as a habitat for people we will also be making it more congenial as a habitat for birds.

Deliberate human interference with birds, on which so much attention has been lavished, is of very limited importance compared with these indirect influences. Some birds, particularly jays, magpies, carrion- and hooded crows, and sparrow-hawks are systematically destroyed over wide areas for game preservation, and by farmers, who also carry on war against woodpigeons and sometimes rooks, sparrows, starlings and other species. Herons, kingfishers, and cormorants are similarly attacked by fishing people on many waters. Several generally harmless birds such as kestrels, little owls and barn-owls are also shot or trapped by the ignorant and prejudiced. It is not clear whether these efforts are at present effective in securing a significant reduction in numbers over the British Isles generally except among the larger birds of prey, which have undoubtedly been very much reduced and are still limited by human destruction. Other birds such as the magpie, carrion-crow, and heron are undoubtedly severely reduced in certain regions by shooting, but are virtually uncontrolled over wide areas, which serve as a perpetual reservoir for replacing casualties.

In considering the justification for attempting preventive destruction of birds several questions have to be answered. The first is whether the birds whose destruction is urged are in fact causing appreciable loss or damage. The second is whether apart from this they are performing some essential or valuable role in other ways. The third is whether and if so how their place would be filled by nature if they were eliminated. The fourth is whether practicable methods and resources are available sufficient to bring about and maintain a worthwhile reduction in numbers in face of the powers of replacement of the species concerned. It is highly questionable on all available evidence whether if these problems are realistically faced a case can be firmly made out for general measures against a single British wild bird in present conditions. Rabbits, to take one comparison, are estimated to exceed thirty millions in the British Isles and to consume several million tons a year of crops and grazing, besides ruining large acreages by eating out all the nourishing herbage and occupying some hundreds of men full time in making, installing and maintaining rabbit fencing, in trapping, and in making good damage in such ways as replacing injured young trees. Yet rats, which destroy vast quantities of food ready for human consumption, spread disease and do great damage on the farm and on game estates, are probably an even greater burden. Compared with either of these even the gross damage inflicted by any wild bird in Britain is a drop in the bucket. It is fair to add, however, that while the economic loss caused by wild birds has often been vastly exaggerated there is no scientific evidence to support the equally large claims for benefits alleged to be given by wild birds in destroying agricultural and garden pests. It would be wise to assume until much more contrary evidence can be brought forward that wild birds do us very little harm and also very little good.

It is not sufficient, however, to take only a national view. Birds may do negligible damage in the country as a whole and yet inflict serious harm on a particular orchard or farm or game estate. Even here the temptation to exaggeration is strong and many losses which cause great excitement at the time make little or no difference to the eventual crop harvested or number of birds reared. Losses from bad weather, disease, and even bad management normally far outweigh losses or damage inflicted by birds except where a grower or rearer on an extremely small scale is unlucky enough to attract to himself an undue proportion of the damage done over some wide area. Where small quantities of crops or stock are exceptionally valuable, vulnerable, and concentrated the difficulty may become acute, but given that national extermination of the birds concerned is neither justifiable nor practicable shooting and trapping are futile or at best uneconomic in most of these cases. If as much money and ingenuity were put into preventing the damage as is devoted to destroying the miscreants the trouble could often be remedied without insuperable difficulty, but the old hunting instinct is strong, and an excuse for lying in wait with a gun is often hard to resist.

There are exceptional cases where protection is so vital and other forms of safeguarding so impracticable that systematic local destruction of a threatening species may prove the only way of achieving the object, as in the case of a breeding-place of some exceedingly rare bird which is threatened by a plague of egg-seeking crows or magpies, or where crows acquire the habit of plucking out lamb's eyes. Such direct repression is however such a burdensome, costly and unending commitment that increased efforts to find practical alternatives are necessary. It seems, in particular, worth investigating how far methods of game-rearing and poultry-rearing and the persecution of large birds of prey may be factors in the modern unwelcome increase of several species of crows and how far revival of suitable predators might discourage them. It would be ironic if efforts to create a paradise for game-chicks swept clean of large birds of prey should prove to be one of the forces which are making Britain a paradise for crows.

Birds are so mobile, so elusive and so adapted to making good high mortality that the more abundant and aggressive species would be difficult to control even if their control were a vital human interest. It is also very difficult to guess what the effect of drastic measures would be, even if they were practicable. For instance, if we exterminated the rook other species would certainly occupy at least part of the ecological niche thus left open, but which species, and with what ultimate effect? Perhaps the black-headed and other gulls, which are already increasing as field birds, would multiply. Perhaps the jackdaw and magpie would take advantage of the opportunity. Perhaps woodpigeons, aided by afforestation, would fill part of the gap. Whatever happened a chain of widespread and unforeseen consequences would be started. If we can find out enough about these processes to justify us in risking drastic interference we may well at the same time find out that direct intervention is a clumsy and unnecessary method of seeking changes which are better secured by indirect means. For example we have good reason to believe that several of our most abundant species have become so owing to specially favourable opportunities presented them by developments in agriculture, forestry, gardening and fisheries. There is no reason why further developments should not reverse the influence partly or entirely, and this is probably already happening with certain species. At the same time, fresh human developments may well give a stimulus and opportunity for expansion to additional species. This again is actually happening as we see with the colonisation of the black redstart and little ringed plover, and the current increase of gulls and waterfowl. So far as we can judge such a tendency would be very desirable. Our civilised development, in its earlier stages, has presented too few species of birds with too great opportunities for expansion, and has enabled them to compete at an undue advantage with others all along the line, thus increasing their dominance still further. The more handicaps can be placed on the few great dominant species, and the more competitive disadvantages of other species can be abated, the richer and better balanced our avifauna is likely to become. When starlings for example become too common they are tempted to look for nesting holes at the expense of woodpeckers on a scale which is trivial in relation to starling numbers but may be quite significant in relation to the nesting success of woodpeckers. The same is true of house-sparrows in relation to house-martins. Aggressive habits which may be unimportant in a stable or declining species may become a serious threat to others in an expanding species.

When we speak of civilisation upsetting the balance of nature we are concerned principally with four main factors—the tendency to reduce or eliminate certain habitats such as marshes and heaths without providing suitable substitutes, the tendency to persecute and even to eliminate certain species which we regard as sporting birds or as pests or vermin, the tendency to introduce alien species, deliberately or by accident, and the tendency to offer undue opportunities for certain species to become dominant. All these tendencies are artificial and it is safest to work on the assumption that all of them are undesirable and should wherever possible be checked and in some cases reversed.

We can check the elimination of scarce habitats by giving them permanent status as nature reserves. We can check by education and by law the excessive killing of wildfowl and wild game and we can discourage the shooting and trapping as pests at any rate of the larger hawks whose restoration is of outstanding importance and ought to be possible over a period. We can by legislation prohibit the uncontrolled introduction of alien species in conditions liable to result in their becoming established in the avifauna. We are not in a position to say definitely that we can check the tendency to encourage undue dominance by certain species, but there is no reason why we should not try, provided that we can acquire enough knowledge of the processes by which dominance has been assisted. For example, just as the numbers of house-sparrows appear to have been reduced as horses have been replaced by motor vehicles which do not scatter corn from nose-bags, so the numbers of rooks might well be reduced by improved methods of harvesting, such as the combine-harvester, resulting in less grain lying about the fields and possibly also by an increased livestock and poultry population competing for gleanings. A reduction in this way is far less improbable than a reduction by emotional campaigns on the part of agricultural committees and farmers. A careful study of factors already at work in certain areas which keep down the numbers of carrion and hooded crows, magpies and rooks, starlings, house-sparrows, and gulls would probably yield useful pointers.

Human innovations will in some cases lead to difficulties with certain birds on account of their special habits. The opening of milk-bottles and tearing of paper by tits is a conspicuous example, but perhaps the worst trouble is the habit certain species have of collecting in flocks on airfields and flying up so that they come into collision with aircraft alighting and taking off. At the very high speeds now becoming prevalent it is extremely hard to construct aircraft so that their wind-screens cannot be smashed or their jet intakes blocked by the impact of a bird, and although the risk may not look very great the number of accidents so caused is quite significant. It is extremely difficult to find any practical way of keeping these birds clear, and we must hope that they will help to solve the problem themselves by learning to avoid jet aircraft as they have on the whole learnt to avoid telegraph wires and even motor-cars. Here again attempts to destroy the birds are futile, as others promptly come in to take their place.

Summing up this review of the birds which live in Britain among people, or on land whose character has been changed by people cultivating or developing it, the conclusion must be that in shaping the landscape of Britain we also shape the bird life, and that as we increase this rate of change in the landscape so we are increasing the rate and the complexity of the effects upon bird life. It also appears that the old conception of a rich primeval bird life being steadily impoverished by civilisation is ceasing to be true, and that civilisation is enriching and diversifying our bird life in some ways while reducing it in others.

A review of the main tangible works of civilisation and of the way in which birds take advantage of them also indicates that more and more species are exploiting our works in more and more ways. A similar picture results from considering separately farmlands, gardens, orchards and hedgerows, and towns and buildings, with the birds attached to each of these groups of habitats, although here it is interesting to find that farmland, traditionally the most hospitable to birds of all man-made landscapes, is being rendered much less so by modern agricultural methods, whereas towns, which have long been the least hospitable, are now beginning to make themselves more attractive.

Consideration of over thirty species of land birds which have become most closely linked with people or their work shows nothing obvious which they have in common. Some are very adaptable, others highly specialised; some migratory, others stay-at-homes; some very tame, others wary and unapproachable; some increasing and spreading, others rapidly declining. Nearly all sorts of birds in fact are capable of living among people.

Such a review naturally leads us to ask what needs to be done to safeguard and enrich our avifauna, and how far we can hope to do it. The answer suggested by the evidence is that there is comparatively little need or justification for direct intervention so far as the birds we are here considering are concerned, although a very few pests must be shot or trapped in certain critical areas, and some very scarce species need to be specially nursed as long as they hang on precariously without being able to secure a sufficiently wide and firm base for assured survival. As by far the greater part of the influence which we exert on bird numbers and distribution is indirect and unconscious, legislation can give only limited results, even if it is soon, as we may hope, revised on much more intelligent lines than at present. Increased study and education, which enlarge our grasp and control of both direct and indirect influences upon birds, are, therefore, most important. Unless the human population should be exterminated the interests of birds must always occupy a subordinate place, except in a few localities where they can be made overriding. Yet the close and growing similarity between the interests of birds and people in the shaping of the landscape encourages a hope that with intelligent and imaginative study Britain can be much improved as a habitat for both.

LIST OF SCIENTIFIC NAMES (AND ALTERNATIVE ENGLISH NAMES) OF BIRDS MENTIONED IN THE TEXT

The following list gives scientific names of birds mentioned in the text. Where the English name used in this book differs from that adopted in Witherby's *Handbook of British Birds* the latter is also shown, in brackets. American names for the same species, where different, are also added in brackets. The order follows the Handbook throughout.

Raven=*Corvus corax* L.
Hooded Crow=*Corvus cornix* L.
Carrion-Crow=*Corvus corone* L.
Rook=*Corvus frugilegus* L.
Jackdaw=*Corvus monedula* L.

Magpie=*Pica pica* (L.)

Jay=*Garrulus glandarius* (L.)

Chough=*Pyrrhocorax pyrrhocorax* (L.)

Starling=*Sturnus vulgaris* L.

Hawfinch=*Coccothraustes coccothraustes* (L.)

American Cardinal = *Richmondena cardinalis* (L.)

Greenfinch=*Chloris chloris* (L.)

Goldfinch=*Carduelis carduelis* (L.)
Siskin=*Carduelis spinus* (L.)
Lesser Redpoll = *Carduelis flammea cabaret* (P.L.S. Müll.)

Twite=*Carduelis flavirostris* (L.)
Linnet=*Carduelis cannabina* (L.)

Serin=*Serinus canarius* (L.)

Bullfinch=*Pyrrhula pyrrhula* (L.)

Crossbill=*Loxia curvirostra* L.

Chaffinch=*Fringilla coelebs* L.
Brambling=*Fringilla montifringilla* L.

Corn-bunting=*Emberiza calandra* L.
Yellowhammer (Yellow Bunting)=*Emberiza citrinella* L.
Cirl-bunting=*Emberiza cirlus* L.
Reed-bunting=*Emberiza schoeniclus* (L.)

Snow-bunting=*Plectrophenax nivalis* (L.)
House-sparrow=*Passer domesticus* (L.)
Tree-sparrow=*Passer montanus* (L.)

Crested lark=*Galerida cristata* (L.)

Woodlark = *Lullula arborea* (L.)

Skylark = *Alauda arvensis* L.

Tree-pipit = *Anthus trivialis* (L.)
Meadow-pipit = *Anthus pratensis* (L.)
Rock-pipit = *Anthus spinoletta* (L.)

Blue-headed wagtail = *Motacilla flava flava* (L.)
Sykes's wagtail = *Motacilla flava beema* (Sykes)
Grey-headed wagtail = *Motacilla flava thunbergi* Billberg
Ashy-headed wagtail = *Motacilla flava cinereocapilla* Savi
Black-headed wagtail = *Motacilla flava feldegg* Michah.
Yellow wagtail = *Motacilla flava flavissima* (Blyth)
Grey wagtail = *Motacilla cinerea* Tunst.
Pied wagtail = *Motacilla alba yarrellii* Gould
White wagtail = *Motacilla alba alba* L.

Treecreeper (Brown creeper in America) = *Certhia familiaris* L.

Nuthatch = *Sitta europaea* L.

Great tit = *Parus major* L.
Blue tit = *Parus caeruleus* L.
Coal tit = *Parus ater* L.
Irish coal tit = *Parus ater hibernicus* O-Grant
Marsh tit = *Parus palustris* L.
Willow tit (Chickadee in America) *Parus atricapillus* L.

Long-tailed tit = *Aegithalos caudatus* (L.)

Bearded tit = *Panurus biarmicus* (L.)

Red-backed shrike = *Lanius collurio* L.

Waxwing = *Bombycilla garrulus* (L.)

Spotted flycatcher = *Muscicapa striata* (Pall.)
Pied flycatcher = *Muscicapa hypoleuca* (Pall.)

Goldcrest = *Regulus regulus* (L.)
Firecrest = *Regulus ignicapillus* (Temm.)

Chiffchaff = *Phylloscopus collybita* (Viell.)
Willow-warbler = *Phylloscopus trochilus* (L.)

Moustached warbler = *Lusciniola melanopogon* (Temm.)

Savi's warbler = *Locustella luscinioides* (Savi)
Grasshopper warbler = *Locustella naevia* (Bodd.)

Icterine warbler = *Hippolais icterina* (Viell.)
Melodious warbler = *Hippolais polyglotta* (Viell.)

Garden warbler = *Sylvia borin* (Bodd.)
Blackcap = *Sylvia atricapilla* (L.)
Whitethroat = *Sylvia communis* (Lath.)
Lesser Whitethroat = *Sylvia curruca* (L.)
Dartford warbler = *Sylvia undata* (Bodd.)

Fieldfare = *Turdus pilaris* L.
Mistlethrush = *Turdus viscivorus* L.
Throstle (Song-thrush) *Turdus ericetorum* Turton

Redwing = *Turdus musicus* L.
Ring-ouzel = *Turdus torquatus* L.
Blackbird = *Turdus merula* L.

Wheatear = *Oenanthe oenanthe* (L.)
Whinchat = *Saxicola rubetra* (L.)
Stonechat = *Saxicola torquata* (L.)

Redstart = *Phoenicurus phoenicurus* (L.)
Black redstart = *Phoenicurus ochrurus* (Gm.)

Nightingale = *Luscinia megarhyncha* Brehm

Robin = *Erithacus rubecula* (L.)

Dunnock (Hedge-sparrow) = *Prunella modularis* (L.)

Wren (Winter-wren in America) = *Troglodytes troglodytes* (L.)
St. Kilda Wren = *Troglodytes troglodytes hirtensis* Seeb.

Swallow (Barn-swallow in America) = *Hirundo rustica* L.

House-martin = *Delichon urbica* (L.)

Sand-martin (Bank-swallow in America) = *Riparia riparia* (L.)

Swift = *Apus apus* (L.)

Nightjar = *Caprimulgus europaeus* L.

American nightjar (Eastern nighthawk) = *Chordeiles minor* (Forster)

Bee-eater = *Merops apiaster* L.

Indian Roller = *Coracias benghalensis* L.

Kingfisher = *Alcedo atthis* (L.)

Green-woodpecker = *Picus viridis* L.
Pied woodpecker (Great spotted woodpecker) = *Dryobates major* (L.)
Barred woodpecker (Lesser spotted woodpecker) = *Dryobates minor* (L.)

Wryneck = *Jynx torquilla* L.

Cuckoo = *Cuculus canorus* L.

Kea = *Nestor notabilis* Gould

Little owl = *Athene noctua* (Scop.)
Long-eared owl = *Asio otus* (L.)

Tawny owl = *Strix aluco* L.

Barn-owl = *Tyto alba* (Scop.)

Peregrine (Duck-hawk in America) = *Falco peregrinus* Tunst.
Hobby = *Falco subbuteo* L.
Kestrel = *Falco tinnunculus* L.

Buzzard = *Buteo buteo* (L.)

Marsh-harrier = *Circus aeruginosus* (L.)

Goshawk = *Accipiter gentilis* (L.)
Sparrowhawk = *Accipiter nisus* (L.)

Kite = *Milvus milvus* (L.)

Honey-buzzard = *Pernis apivorus* (L.)

Turkey vulture or Turkey-buzzard = *Cathartes aura* (L.)

Spoonbill=*Platalea leucorodia* L.

Common heron=*Ardea cinerea* L.

Bittern=*Botaurus stellaris* (L.)
Mute swan=*Cygnus olor* (gm.)
Grey geese=*Anser* spp.

Mallard=*Anas platyrhyncha* L.
Wigeon=*Anas penelope* L.

Pochard=*Aythya ferina* (L.)
Tufted duck=*Aythya fuligula* (L.)
Scaup=*Aythya marila* (L.)
Smew=*Mergus albellus* L.

Cormorant=*Phalacrocorax carbo* (L.)

Gannet=*Sula bassana* (L.)

Storm-petrel=*Hydrobates pelagicus* (L.)
Manx shearwater=*Puffinus puffinus* (Brünn.)

Fulmar=*Fulmarus glacialis* (L.)
Great crested grebe=*Podiceps cristatus* (L.)
Dabchick (Little grebe)=*Podiceps ruficollis* (Pall.)

Woodpigeon=*Columba palumbus* L.
Stock-dove=*Columba oenas* L.
Rock-dove=*Columba livia* Gm.

Turtle-dove=*Streptopelia turtur* (L.)

Black-tailed godwit=*Limosa limosa* (L.)

Curlew=*Numenius arquata* (L.)

Snipe=*Capella gallinago* (L.)

Turnstone=*Arenaria interpres* (L.)

Purple sandpiper=*Calidris maritima* (Brünn.)

Ruff (and reeve)=*Philomachus pugnax* (L.)

Redshank=*Tringa totanus* (L.)

Ringed plover=*Charadrius hiaticula* L.

Little ringed plover=*Charadrius dubius* Scop.

Golden plover=*Pluvialis apricaria* (L.)

Lapwing=*Vanellus vanellus* (L.)

Black-winged stilt=*Himantopus himantopus* (L.)

Avocet=*Recurvirostra avosetta* L.

Oystercatcher=*Haematopus ostralegus* L.

Stone-curlew=*Burhinus oedicnemus* (L.)

Great bustard=*Otis tarda* L.

Common crane=*Grus grus* (L.)

Black tern=*Chlidonias niger* (L.)

Gull-billed tern=*Gelochelidon nilotica* (Gm.)

Arctic tern=*Sterna macrura* Naumann

Black-headed gull=*Larus ridibundus* L.

Mew gull (Common gull)=*Larus canus* L.

Herring gull=*Larus argentatus* Pont.

Lesser blackback (Lesser black-backed gull)=*Larus fuscus* L.

Great blackback (Greater black-backed gull)=*Larus marinus* L.

Kittiwake=*Rissa tridactyla* (L.)

Great skua=*Stercorarius skua* (Brünn.)

Corncrake=*Crex crex* (L.)

Water-rail=*Rallus aquaticus* L.

Moorhen (Florida gallinule in America)=*Gallinula chloropus* (L.)

Coot=*Fulica atra* L.

Red grouse=*Lagopus scoticus* (Lath.)

Pheasant=*Phasianus colchicus* L.

Common partridge (Hungarian partridge in America)=*Perdix perdix* (L.)

Red-legged partridge=*Alectoris rufa* (L.)

Quail=*Coturnix coturnix* (L.)

BIBLIOGRAPHICAL NOTES

Chapter 1

General picture of influence of men on natural features in Britain:

CLAPHAM, SIR JOHN (1949). *A Concise Economic History of Britain from the Earliest Times to 1750.* Cambridge, University Press.

CLARK, GRAHAME (1937). *Prehistoric England.* London, Batsford.

TANSLEY, A. G. (1949). *Britain's Green Mantle.* London, Allen & Unwin.

Review of past changes in bird life in Britain:

NICHOLSON, E. M. (1926). *Birds in England.* London, Chapman & Hall, pp. 1-119.

Recent changes in status:

ALEXANDER, H. G. (1949). Bearded Tits at Cley, Norfolk. *Brit. Birds, 42:* 289-91.

BROWN, P. E. & LYNN-ALLEN, E. (1948). Breeding of Avocets in England in 1947. *Brit. Birds, 41;* 14-17.

COOKE, R. (1946). Black Terns Breeding in Sussex. *Brit. Birds, 39;* 71-72.

STATON, J. (1945). The Breeding of Black-winged Stilts in Nottinghamshire in 1945. *Brit. Birds, 38:* 322-28.

HINDE, R. A. & THOM, A. S. (1947). The Breeding of the Moustached Warbler in Cambridgeshire. *Brit. Birds, 40:* 98-104.

Density of Woodland Birds:

LACK, D. (1937). A review of Bird Census Work and Population Problems. *Ibis,* ser. 14, *1:* 369-95 (p. 372).

Chapter 2.

Robin nesting on bed:

ALEXANDER, H. G. (1929). Robin Nesting on a Bed in an Occupied Room. *Brit. Birds, 23:* 37-38.

Tits and milk bottles:

FISHER, JAMES & HINDE, R. A. The opening of Milk Bottles by Birds. *Brit. Birds, 42:* 347-57.

Little Ringed Plover:

PARRINDER, E. R. (1948). The Little Ringed Plover in the London Area in 1947. *Brit. Birds, 41:* 41-45.

PARRINDER, E. R. (1948D). The Little Ringed Plover in Southern England in 1948. *Brit. Birds, 41:* 377-83.

CHISLETT, R., DAWSON, K. & ALLISON, F. R. (1948). The Little Ringed Plover in Yorkshire. *Brit. Birds, 41:* 384-85.

Duke of Wellington and sparrows in 1851 Exhibition:

HOBHOUSE, C. (1937). *1851 and the Crystal Palace.* London, John Murray, p. 60.

Chapter 3.

General outline of British farming:

SCOTT WATSON, J. A. (1938). *The Farming Year.* London, Longmans.

Intensive study of an Oxfordshire Farm:

ALEXANDER, W. B. (1932). Bird Population of an Oxfordshire Farm. *J. Anim. Ecol. 1:* 58-64.

CHAPMAN, W. M. M. (1939). Bird Population of an Oxfordshire Farm. *J. Anim. Ecol. 8:* 286-99.

New black-headed gullery near London:

HURCOMB, SIR C. (1947). Breeding Colony of Black-headed gulls in Middlesex. *Brit. Birds, 40:* 28-29.

Use of name, "Mew gull":

ALEXANDER, W. B. (1928). *Birds of the Ocean.* New York, Putnams, p. 131.

Chapter 4.

Rooks on London Buildings:

GLEGG, W. E. (1935). *A History of the Birds of Middlesex.* London, Witherby, p. 8.

Rook Investigation 1944-46:

AGRICULTURAL RESEARCH COUNCIL (1948). Report Received by the Ministry. *Agriculture* (Journal of the Ministry of Agriculture), *55:* 20-23, and unpublished data.

German rookeries:

RORIG, G. (1900). Die Verbreitung der Saatkrähe in Deutschland. *Arb. biol. Anst. Land-Forstwirt, Berlin, 1:* 271-84.

Dr. Collinge's food investigation:

COLLINGE, W. E. (1924). *The Food of Some British Wild Birds: A study in Economic Ornithology.* York, publ. by Author.

Rook population of the Upper Thames:

NICHOLSON, E. M. & B. D. (1930). The Rookeries of the Oxford District: A Preliminary Report. *J. Ecol. 18:* 51-66.

ALEXANDER, W. B. (1933). The Rook Population of the Upper Thames Region. *J. Anim. Ecol. 2:* 24-35.

Rooks and gapeworms:

ELTON, C. & BUCKLAND, F. (1928). The Gapeworm (*Syngamus trachea* Montagu) in Rooks (*Corvus frugilegus, L.*), *Parasitology, 20:* 448-50.

Parasites in Rook's Nest:

BLAIR, K. G. (1931). A note on the Arthropod Fauna of Rooks' Corn-bunting census. Nests. *Ent. Mon. Mag. 67:* 35-36.

COURSE, H. A. (1941). Some Census Work on the Corn-bunting. *Brit. Birds, 35:* 154-55.

Destruction of Skylarks:

KNOX, A. E. (1849). Ornithological Rambles in Sussex. London, Van Voorst, pp. 126-32.

Yellow wagtail mutants:

TUCKER, B. W. (1949). Species and Subspecies. *Brit. Birds, 42:* 000-00 (p. 193).

Little Owl's food and economic status:

HIBBERT-WARE, A. (1938). *Report of the Little Owl Food Enquiry 1936-37.* London, Witherby (offprinted from *British Birds*).

Barn Owl numbers and distribution:

BLAKER, G. B. (1934). *The Barn Owl in England and Wales.* London, Royal Society for the Protection of Birds (offprinted from *Bird Notes and News*).

Lapwing habitats:

NICHOLSON, E. M. (1938-39). Report on the Lapwing Habitat Inquiry, 1937. *Brit. Birds, 32:* 170-91, 207-29, 255-59.

LISTER, M. D. (1939). An Account of the Lapwing Population on a Surrey Farm. *Brit. Birds, 32:* 260-71.

Lapwing territories:

HOWARD, H. ELIOT (1920). *Territory in Bird Life.* London, Murray, pp. 58-61.

Transatlantic flight:

WITHERBY, H. F. (1928). A Transatlantic Passage of Lapwings. *Brit. Birds, 22:* 6-12.

Corncrake Report:

NORRIS, C. A. (1945). Summary of a Report on the Distribution of the Corn Crake. *Brit. Birds, 38:* 142-48, 162-68.

Corncrake display:

MASON, A. G. (1945). The Display of the Corn Crake. *Brit. Birds, 38:* 351-52.

Partridge population:

MIDDLETON, A. D. (1936). Factors Controlling the Population of the Partridge (*Perdix perdix*) in Great Britain. *Proc. Zool. Soc. Lond., 1935:* 795-815.

Red-legged partridge introductions:

GLADSTONE, H. S. (1930). *Record Bags and Shooting Records.* London, Witherby, 2nd ed., pp. 72-74.

KIRKMAN, F. B. and HUTCHINSON, H. G. (1924). *British Sporting Birds.* London, Jack, pp. 36-37.

MONTAGU, G. (1866). *A Dictionary of British Birds.* London, Van Voorst, ed. Edward Newman, pp. 229-30.

KNOX, A. E. (1849). *Ornithological Rambles in Sussex.* London, Van Voorst, pp. 168-69.

Quail, slaughter in Mediterranean Countries:

HENNICKE, C. R. (1912). *Handbuch des Vogelschutzes.* Magdeburg, Kreutzsche.

GLADSTONE, H. S. (1930). *Record Bags and Shooting Records.* London, Witherby, 2nd ed., pp. 75-78.

Quail, journey from Italy to Sweden traced by ringing:

LONNBERG, E. (1938). Notes on the Migration of Swedish Birds. *Proc. VIII Int. Orn. Congr. Oxford, 1934:* 602-19 (p. 619).

Recent distribution trends in Britain:

TUCKER, B. W. (1945). Report on Quail in the British Isles in the Summer of 1944. *Brit. Birds, 38:* 250-52.

MOREAU, R. E. (1949). Quail in Oxfordshire. *Brit. Birds, 42:* 191-92.

Chapter 5.

Estimate of cultivated area of world:

RUSSELL, SIR J. (1949). World Population and World Food Supplies. *Nature, 164:* 379.

Estimates of cat population:

FITTER, R. S. R. (1949). *London's Birds.* London, Collins *New Naturalist,* p. 199.

EVENING STANDARD (1949). Reporting Mrs. M. Larwood, Registrar of the National Animal Registration Service in issue 15 Nov.

Acreage under fruit:

MINISTRY OF AGRICULTURE AND FISHERIES. *Fruit Census 1944, England and Wales.* London.

Shrubs recommended for planting to attract birds:

COMMITTEE ON BIRD SANCTUARIES IN THE ROYAL PARKS (1948). *Birds in London.* London, H.M. Stationery Office, p. 3.

Greenfinch revisiting trap:

NICHOLSON, E. M. & WILLSON, M. W. (1928). The Oxford Trapping Station. *Brit. Birds, 21:* 292-93.

Goldfinch in London area:

HARTING, J. E. (1866). *Birds of Middlesex.* London, Van Voorst, p. 86.

GLEGG, W. E. (1935). *A History of the Birds of Middlesex.* London, Witherby, pp. 25-27.

Yellowhammer farm population:

CHAPMAN, W. M. M. (1939). Bird Population of an Oxfordshire Farm. *J. Anim. Ecol. 8:* 286-99 (p. 298).

Relative numbers of throstles and blackbirds:
LACK, D. & VENABLES, L. S. V. (1939). Habitat Distribution of British Woodland Birds. *J. Anim. Ecol. 8:* 39-71 (p. 49).

Analysis of British Trust for Ornithology's Nest Records of Throstle:
SILVA, E. T. (1949). Nest Records of the Song Thrush. *Brit. Birds, 42:* 97-111.

Blackbirds, communal activities:
MORLEY, A. (1937). Some activities of Resident Blackbirds in Winter. *Brit. Birds, 31:* 34-41.

Blackbird mortality calculated from ringing returns of the British Trust for Ornithology:
LACK, D. (1943). The Age of the Blackbird. *Brit. Birds, 36:* 166-75.

Robin, territory and mortality:
LACK, D. (1943). *The Life of the Robin.* London, Witherby.
BROOKS-KING, M. (1944). Some Observations on a Tame Robin. *Brit. Birds, 38:* 130-32.

Sexual behaviour of Dunnock:
SELOUS, E. (1933). *Evolution of Habit in Birds.* London, Constable, p. 109.

Polygamy in wrens:
KLUYVER, H. N. & Others (1940). De levenswijze van der Winterkoning. *Limosa, 13:* 1-51.

Wren song:
ARMSTRONG, E. A. (1944). Some notes on the Song of the Wren. *Brit. Birds, 38:* 70-72.

Wrens' roosting habits:
DUNSHEATH, M. H. & DONCASTER, C. C. (1941). Some Observations on Roosting Birds. *Brit. Birds, 35:* 145-47.

Wryneck in Kensington Gardens in 19th Century:
HAMILTON, E. (1879). The Birds of London. *Zoologist, 1879:* 273-91.

Chapter 8.

Jackdaw society:
LORENZ, K. (1938) A Contribution to the Comparative Sociology of Colonial-nesting Birds. *Proc. VIII Int. Orn. Congr. Oxford, 1934:* 207-18.

Starling, population and roosting:

MARPLES, B. J. (1934). The Winter Starling Roosts of Great Britain, 1932-1933. *J. Anim. Ecol. 3:* 187-203.

WYNNE-EDWARDS, V. C. (1929). The Behaviour of Starlings in Winter. *Brit. Birds, 23:* 138-53, 170-80.

Spread of starlings in America:

KALMBACH, E. R. (1928). The European Starling. *Bull. U.S. Dept. Agric.* No. 1571.

Distribution in New Zealand:

MARPLES, B. J. (1946). List of the Birds of New Zealand. *Bull. Orn. Soc. N.Z. Dunedin.* March 1946, Supplement vii.

Roosting in London:

FITTER, R. S. R. (1943). The Starling Roosts of the London Area. *London Naturalist, 1942:* 3-23.

Increase on St. Kilda:

FISHER, J. (1948). St. Kilda a Natural Experiment. *New Naturalist Journal 1:* 91-108 (p. 107).

Food and Economic status:

KLUYVER, H. N. (1938). The Importance of the Starling in Agriculture. *Proc. VIII Int. Orn. Congr. Oxford, 1934:* 720-25.

House-sparrow, general review of world distribution trends and of food:

SOUTHERN, H. N. (1945). The Economic Importance of the House-Sparrow, *Passer domesticus* L: a Review. *Annals of Applied Biology, 32:* 57-67.

Sparrow population of Kensington Gardens:

NICHOLSON, E. M. (1927). A Bird Census of Kensington Gardens. *Discovery, 7:* 281.

Black Redstarts on Pett Level since 1909:

COOKE, R. (1948). Black Redstarts on Pett Level, East Sussex. *Brit. Birds, 41:* 46-48.

Spread of Black Redstarts in Britain:

WITHERBY, H. F. & FITTER, R. S. R. (1942). Black Redstarts in England in the Summer of 1942. *Brit. Birds, 36:* 132-39; and subsequent papers by FITTER (listed in *Brit. Birds, 39:* 211 up to 1946).

Swallow population, reproduction and parasites:

BOYD, A. W. (1935). Report on the Swallow Enquiry 1934. *Brit. Birds, 29:* 3-21.

(1935D). A Fly Parasitic on the Swallow. *Brit. Birds, 28:* 225.

(1936). Report on the Swallow Enquiry 1935. *Brit. Birds, 30:* 98-116.

Swift migration:

HURRELL, H. G. (1948). Simultaneous Watch for Migrant Swifts, May 11th, 1947. *Brit. Birds, 41:* 138-45.

Information derived from the following sources is not separately cited:

WITHERBY, H. F. and Others (1938-41). *The Handbook of British Birds.* London, Witherby. 5 Vols.

NICHOLSON, E. M. and KOCH, L. (1936). *Songs of Wild Birds.* London, Witherby.

NICHOLSON, E. M. and KOCH, L. (1937). *More Songs of Wild Birds.* London, Witherby.

NICHOLSON, E. M. (1929). *How Birds Live.* London, Williams and Norgate, 2nd Ed., for references to bird censuses and flight speeds.

NICHOLSON, E. M. (1927). A Bird Census in the French Alps. *Discovery, 8:* 357-59, 400-04.

HARRISSON, T. H. (1931). On the Normal Flight Speeds of Birds. *Brit. Birds, 25:* 86-96.

ROBERTS, B. B. (1931). On the Normal Flight Speed of Birds. *Brit. Birds, 25:* 220-22.

MACDONALD, J. D. (1949). *Birds of Britain.* London, Bell, pp. 11, 147-48 (bird weights).

and also, as regards records of bird weights:

GLADSTONE, H. S. (1930). *Record Bags and Shooting Records.* London, Witherby, 2nd Ed.

Local records not attributed to the author's own observation are based upon the regular bird reports published for the area concerned. Particular references to distribution in Switzerland are based on:

MEYLAN, O. and HALLER, W. (1946). Artliste der Schweizerischen Vogel. *Neue Aargauer Zeitung* for *Zentralverband Schweizerischer Vogelschutzvereine*, Aarau and

GUGGISBERG, C. A. W. (undated) *Vogel der Schweiz.* Bern. Hallwag. 2 vols.

Both these works are recommended as models of their kind, the second being extremely well illustrated.

INDEX OF BIRD NAMES

Separate sections devoted to a species are referenced in heavy type.

GENERAL INDEX